No Grid Survival Proj

[10 Books in 1] The Comprehensive DIY Projects to Become Self-Sufficient in Energy, Water, Food Supply, Housing Solutions, and Emergency Preparedness

Roderick Wade

Table of Contents

FREE SUPPLEMENTARY RESOURCES

Scan the QR code below to download the eBooks:

- ✔ Simple and Functional DIY Furniture
- ✔ Traps for Fishing and Hunting
- ✔ DIY Miniature Greenhouse
- ✔ Shelter Building Techniques
- ✔ Rainwater Collection Device
- ✔ Low-Cost Water Filtration System
- ✔ DIY Miniature Greenhouse
- ✔ DIY Wind Turbine for Home Use
- ✔ Food Preservation Without Electricity
- ✔ Off-Grid Cooking Recipes
- ✔ DIY Navigation Tools
- ✔ DIY Survival Kit
- ✔ Natural First Aid Kit

SCAN THE QR CODE TO DOWNLOAD

COMPLEMENTARY RESOURCES

Preface

Welcome to a journey less traveled, one that promises not only a return to the essentials but an exploration into a lifestyle that redefines what it means to live meaningfully. This book, "No Grid Survival Projects," invites you into the world of off-grid living —a way of life that embraces independence from the municipal and commercial systems that so many take for granted.

Introduction to the Concept of Off-Grid Living

Imagine a life where every sunrise brings a day not dictated by digital notifications or the hum of electricity but by the rhythm of nature itself. Off-grid living strips away the conventional dependencies on the electrical grid, public water supplies, and centralized resources, challenging you to forge self-sufficiency in every aspect of daily life.

Envision waking up in a home that you've built with your own hands, nestled in a landscape where every plant and every drop of water counts. Your home is powered by the sun and wind, and your days are spent nurturing the garden that provides your meals. This lifestyle isn't a retreat from modernity but an intelligent, conscious response to it—a statement of living sustainably and with purpose.

Off-grid living is about creating a life where you are not only aware of your consumption but actively managing it to harmonize with the surrounding environment. It's about resilience, reducing your ecological footprint, and enhancing your well-being through a profound connection with the earth.

This book isn't just a guide; it's an invitation to think differently about how you live and to challenge the norms that society dictates. It's for those who feel a pull towards self-reliance, for adventurers ready to commit to a lifestyle that requires ingenuity, patience, and a deep respect for the natural world.

As you turn these pages, you'll uncover the essential skills, knowledge, and attitudes needed to successfully navigate and thrive in an off-grid setting. From water collection and food preservation to building your own home and generating energy, each chapter builds on the last, providing a comprehensive framework for a life untethered to the conventional grid.

Ready to redefine what it means to live independently? Let's embark on this enlightening journey together.

Why Your Support Matters for This Book:

Creating this book has been an unexpectedly tough journey, more so than even the most complex coding sessions. For the first time, I've faced the daunting challenge of writer's block. While I understand the subject matter, translating it into clear, logical, and engaging writing is another matter altogether.

Moreover, my choice to bypass traditional publishers has led me to embrace the role of an 'independent author.' This path has had its hurdles, yet my commitment to helping others remains strong.

This is why your feedback on Amazon would be incredibly valuable. Your thoughts and opinions not only matter greatly to me, but they also play a crucial role in spreading the word about this book. Here's what I suggest:

1. If you haven't done so already, scan the QR code at the beginning of the book to download the FREE SUPPLEMENTARY RESOURCES.

2. Scan the QR code below and quickly leave feedback on Amazon!

The optimal approach? Consider making a brief video to share your impressions of the book! If that's a bit much, don't worry at all. Just leaving a feedback and including a few photos of the book would be fantastic too!

Note: There's no obligation whatsoever, but it would be immensely valued!

I'm thrilled to embark on this journey with you. Are you prepared to delve in?
Enjoy your reading!

Book 1: Preparing for Off-Grid Life

Assessing Your Needs and Possibilities

Embarking on an off-grid lifestyle begins with a thorough and honest assessment of your needs and the possibilities available to you. This crucial first step is about understanding both what is essential for your survival and what is feasible given your resources and the environment in which you plan to live.

Understanding Your Basic Needs

The foundation of off-grid living is identifying and securing your basic needs: shelter, water, food, and energy. These are critical components that dictate your quality of life and survival in an off-grid setting.

Shelter provides protection from the elements and a base for your daily activities. It should be sturdy, well-insulated, and suitable for the climate of your chosen location. Think about the materials that will be both sustainable and efficient in maintaining a comfortable indoor environment.

Water is perhaps the most critical resource for human survival. Assessing how you will secure a clean and sustainable water source is paramount. Whether it's through wells, rainwater collection, or nearby natural sources, ensuring that you have reliable access to water and a system for purifying it is essential.

Food security involves planning how you will obtain and store food. This might mean setting up a garden, considering the soil quality and climate, or it could involve raising livestock. You'll need to think about the space, tools, and skills required to sustain these food sources.

Energy needs often go hand-in-hand with your choice of location and the available natural resources. Will you use solar panels, wind turbines, or perhaps a micro-hydro system? Assessing the feasibility of these technologies in your environment will dictate your approach to energy self-sufficiency.

Evaluating Skills and Knowledge

Living off-grid requires a diverse set of skills and a willingness to learn. From building structures to managing a garden, your ability to sustain an off-grid lifestyle largely depends on your skills and knowledge.

Building skills are essential not only for constructing your home but also for maintaining it. Understanding basic carpentry, plumbing, and electrical work can save you significant amounts of money and help keep your home functional and safe.

Agricultural knowledge is crucial if you plan to grow your own food. This includes understanding the types of crops suited to your climate, soil preparation, pest management, and the best practices for harvesting and storing produce.

Water resource management involves not only securing a water supply but also efficiently using and reusing water. Skills in setting up irrigation systems, rainwater catchments, and perhaps even greywater recycling systems will be vital.

Energy management is another critical area. Knowing how to install and maintain renewable energy systems will enable you to manage your power usage effectively. This includes understanding the basics of electrical systems, how to optimize energy consumption, and troubleshooting common issues with renewable energy setups.

Analyzing the Physical Environment

The environment where you choose to live plays a crucial role in what's possible for your off-grid life. Each environment offers unique resources and presents different challenges.

Topography and climate can greatly affect building conditions, agricultural possibilities, and energy solutions. For instance, a mountainous area might offer potential for hydroelectric energy but make building and farming more challenging. Conversely, a flat, sunny area might be ideal for solar panels and agriculture but require significant water transport and management solutions.

Resource availability such as local flora and fauna, soil type, and water sources must be assessed for their practical use. Wooded areas can provide building materials and biomass for energy, but may also have limitations like poor soil for farming or dampness affecting solar panel efficiency.

Renewable energy potential is another key consideration. The availability of sunlight, wind, or water flow can determine the best type of renewable energy system to install. Understanding these elements will help you decide how to invest in and structure your energy solutions.

Resource Availability

Natural and Local Resources The first consideration in off-grid living is identifying the natural resources available at your chosen location. This includes water sources, soil fertility, climatic conditions for solar or wind power, and local flora and fauna that can be utilized sustainably. Access to these resources will significantly influence your daily living and how you plan your food production, water systems, and energy solutions.

However, resources extend beyond the natural. Local markets and community resources play a crucial role in off-grid living. For instance, proximity to a local market can ease the transition by providing access to supplies that cannot be produced independently at the start, such as grains, fabrics, or even hardware. Community resources like shared workshops, knowledge exchange forums, and cooperative farming can offer support and resilience to your off-grid setup.

Online Materials and Tools In today's digital age, online resources are invaluable. Platforms that offer tutorials, forums for community advice, and online stores for purchasing specific tools and technologies that facilitate off-grid living are crucial. These resources can help bridge the gap in your knowledge and provide access to materials and innovations that enhance self-sufficiency, such as tutorials on solar panel installation or sources for buying heirloom seeds.

Legal and Regulatory Considerations

Before you commit to an off-grid location, a thorough understanding of the local legal and regulatory environment is imperative. Laws and regulations regarding land use can affect everything from the type of structure you can build to the methods you can employ for sewage treatment.

Building Codes and Permits Many regions have strict guidelines on construction, which may include the necessity for building permits, adherence to safety standards, and even restrictions on the types of materials used. Ensuring your off-grid home complies with these regulations is essential to avoid legal complications and fines.

Resource Use Regulations Regulations surrounding the use of natural resources such as water and timber are also significant. Some areas might have restrictions on rainwater collection and require specific treatments for using natural water sources. Similarly, cutting down trees for building or firewood might be regulated to prevent environmental degradation.

Off-Grid Waste Management Legal considerations also extend to waste management. Composting toilets, greywater systems, and other off-grid sewage solutions must often meet specific health and environmental standards. Understanding and adhering to these regulations is crucial for a sustainable and legally compliant off-grid life.

Long-term Sustainability

The sustainability of off-grid living is not just about environmental impact but also involves economic and personal factors. It requires assessing the ongoing costs, labor input, and overall impact on your quality of life.

Economic Considerations Consider the initial investment for setting up an off-grid home, including the cost of land, building materials, renewable energy systems, and water management infrastructure. Ongoing expenses such as maintenance of equipment, replacement of tools, and possibly property taxes must also be planned for.

Labor and Time Investment The amount of labor required to maintain an off-grid lifestyle can be significant. From daily tasks such as gardening and animal care to periodic needs like home repairs and energy system maintenance, the labor input can impact your lifestyle choices and personal time.

Environmental Impact Sustainable off-grid living should ideally reduce your ecological footprint. This involves using renewable resources, minimizing waste, and managing land and water in a way that promotes biodiversity and ecological health.

Physical and Mental Preparedness Living off-grid demands not only physical resilience but also mental preparedness. It requires a mindset shift from dependence on conventional services to a more autonomous, self-reliant approach. Emotional readiness and the ability to adapt to unexpected challenges are crucial components of sustainable off-grid living.

In conclusion, transitioning to an off-grid lifestyle necessitates a comprehensive evaluation of your needs, resource availability, and the legal landscape, paired with an unwavering commitment to sustainability. By methodically assessing your basic needs for shelter, water, food, and energy, and aligning these with the skills required and the resources accessible, you can ensure a robust foundation for your off-grid journey. Equally important is understanding and navigating the legal and regulatory frameworks that govern your chosen location to ensure your setup is compliant and sustainable. Lastly, the commitment to off-grid living is not just a logistical shift but a profound lifestyle change that demands both physical resilience and mental fortitude. It's a path that offers immense satisfaction through self-sufficiency but requires a deep dedication to continuous learning and adaptation. Prepare thoroughly, embrace the challenges, and your off-grid life can be both successful and fulfilling.

Planning and Budgeting

Planning and Budgeting for Off-Grid Living

Embarking on an off-grid life is not just a choice; it's an adventure and transformative endeavor that shifts how you interact with the world around you. However, to ensure that this shift is not only successful but also sustainable, meticulous planning and careful budgeting are crucial. The aim is to make your transition feasible, keeping you prepared for both the expected and the unexpected. Here is an in-depth guide on how to effectively plan and budget for your off-grid lifestyle.

Setting Clear Objectives

The first step in planning for an off-grid life is to define what it means for you personally. This involves establishing clear, tangible goals that reflect your desires and realities. Are you aiming to be completely self-sufficient, or do you envision a semi-sustainable lifestyle? Do you plan to live off-grid permanently, or are you looking for a seasonal retreat from modern life? By answering these questions, you can tailor your planning and budgeting to fit your specific needs rather than a one-size-fits-all solution. Clear objectives guide every decision you make, from the location you choose to the type of resources you will rely on.

Research and Resource Evaluation

Successful off-grid living relies heavily on your understanding of the environment and the resources available to you. Begin by conducting thorough research into the various aspects of off-grid living that will affect your setup. This includes:

- **Renewable Energy Options:** Assess the viability of solar, wind, and hydro energy solutions in your intended location. Each has its installation and maintenance costs, and not all will be suitable for every environment.

- **Water and Waste Management:** Investigate your options for managing these critical systems. This could mean drilling a well, setting up rainwater catchments, or installing composting toilets and greywater treatment systems.

- **Local Resources and Labor:** Consider the availability of local building materials, which can reduce costs. Also, understand the labor market in the area if you plan to hire help for setup or ongoing maintenance.

This phase should also include a review of local community resources, which can provide support or share valuable off-grid living knowledge.

Detailed Cost Estimation

Once your goals are set and your research is complete, the next step is to create a detailed cost estimation. This breaks down into three main categories:

1. **Initial Setup Costs:**

   o **Land Acquisition:** If not already owned, purchasing land is often the first and most significant expense.

   o **Home Construction or Modification:** Costs here can vary widely depending on whether you are building from scratch, renovating, or adapting existing structures to suit off-grid needs.

   o **Energy Systems:** Upfront investment in renewable energy systems such as solar panels or wind turbines.

   o **Water Systems:** Includes the cost of well drilling, purchasing tanks for rainwater collection, and setting up filtration systems.

   o **Waste Management:** Installation of non-traditional waste systems like composting toilets and septic tanks.

2. **Recurring Expenses:**

   o **System Maintenance:** Regular upkeep of energy, water, and waste systems to ensure efficiency and longevity.

   o **Property Taxes:** These continue regardless of your off-grid status and must be factored into your budget.

   o **Transportation Costs:** Especially pertinent if your site is remote, these costs include fuel and vehicle maintenance.

   o **Agricultural Inputs:** If you're growing your own food, costs for seeds, tools, and possibly livestock need consideration.

3. **Emergency Funds:**

   o **Unexpected Repairs:** Off-grid systems can sometimes fail; having funds set aside for quick repairs is crucial.

   o **Emergencies:** Whether it's a medical emergency, a failed crop, or another unexpected expense, having a financial cushion can help you manage without stress.

Budget Allocation

Once you have a clear understanding of the various costs involved in transitioning to off-grid living, the next step is to allocate your budget effectively. This process is crucial to ensure that you are prioritizing essential systems and investments that will provide the greatest sustainability and return on investment. Here's a detailed guide on how to strategically allocate your budget for an off-grid lifestyle.

Prioritizing Essential Systems

When allocating your budget, it's important to prioritize investments that will ensure your basic needs are met in the most sustainable and efficient way possible. Here are some key areas to consider:

1. **Energy Systems**:
   - **Solar Power:** Investing in a high-quality solar setup might be expensive initially, but it can provide long-term savings and energy security. Consider the cost of solar panels, inverters, batteries, and installation. High-efficiency panels and reliable storage solutions will ensure that you have a consistent power supply.
   - **Wind Turbines:** In areas with sufficient wind, turbines can complement solar power. Budget for the turbine itself, installation, and maintenance.
   - **Hydropower:** If you have access to flowing water, micro-hydro systems can be an excellent source of energy. Include costs for the turbine, pipes, and installation.

2. **Water Systems**:
   - **Well Drilling and Maintenance:** If you plan to use groundwater, allocate funds for drilling a well and installing a pump system.
   - **Rainwater Harvesting:** Budget for gutters, tanks, and filtration systems to collect and purify rainwater.
   - **Water Purification:** Consider costs for filtration units, UV purifiers, or other systems to ensure your water is safe for drinking.

3. **Shelter and Construction**:
   - **Building Materials:** Allocate funds for sustainable and durable materials for constructing or retrofitting your home.
   - **Insulation and Weatherproofing:** Proper insulation and weatherproofing are critical for maintaining a comfortable living environment and reducing energy consumption.
   - **Labor Costs:** If you're hiring professionals for construction, include labor costs in your budget.

4. **Food Production**:
   - **Gardening Supplies:** Budget for seeds, soil amendments, tools, and possibly greenhouse materials if you're planning year-round food production.
   - **Livestock:** Include costs for purchasing animals, building shelters, feed, and veterinary care.
   - **Preservation Equipment:** Allocate funds for canning supplies, dehydrators, or freezers powered by renewable energy.

Funding and Financing

Exploring various funding options can help mitigate the initial financial burden of transitioning to off-grid living. Here are some strategies to consider:

1. **Personal Savings**:
   - Using your savings can be the most straightforward way to fund your off-grid project. Ensure you have a clear plan and budget to avoid depleting your savings.

2. **Loans**:
   - Look into loans specifically designed for renewable energy installations or sustainable building projects. Many banks and credit unions offer favorable terms for such loans.
   - Home equity loans can also be an option if you own property and need additional funds.

3. **Grants and Incentives**:
   - Research grants available for renewable energy projects, sustainable agriculture, or eco-friendly construction. Various government programs and non-profits offer financial assistance.
   - Tax incentives and rebates are often available for installing renewable energy systems. Check federal, state, and local programs that can reduce your overall costs.

4. **Community Funding**:
   - Crowdfunding platforms can be effective for raising funds for specific projects. Engage your community by sharing your vision and goals.
   - Community-supported agriculture (CSA) programs can also provide funding through advance sales of produce and products.

Contingency Planning

A crucial aspect of budgeting for an off-grid lifestyle is to include a contingency plan. This ensures you are prepared for unexpected costs and setbacks. Here's how to create an effective contingency plan:

1. **Allocate a Percentage**:
   - It is generally recommended to set aside 10-20% of your total budget as a contingency fund. The exact percentage can vary depending on the complexity and scale of your project.

2. **Identify Potential Risks**:
   - List potential risks and unexpected expenses, such as equipment failures, natural disasters, or price increases in materials. Understanding these risks helps in better estimating the required contingency funds.

3. **Accessible Funds**:
   - Ensure that the contingency funds are easily accessible. This might mean keeping a portion of your budget in a liquid form, such as in a savings account, so you can quickly address emergencies.

Regular Review and Adjustment

Once your off-grid life is underway, it's essential to regularly review and adjust your budget based on actual expenses and experiences. Off-grid living often presents unforeseen challenges, and your budget needs to be flexible to accommodate these changes. Here are steps to effectively manage this:

1. **Monthly Reviews**:

    o   Conduct monthly reviews of your expenses versus your budget. Track where your money is going and compare it to your initial estimates.

2. **Adjustments Based on Experience**:

    o   Adjust your budget based on actual expenses. If certain costs are higher or lower than expected, modify your allocations accordingly.

    o   Learn from your experiences. If you find more efficient ways to manage resources or identify areas where you can cut costs, incorporate these adjustments into your budget.

3. **Future Planning**:

    o   Use your budget reviews to plan for future needs. For example, if you anticipate needing new equipment or additional resources in the coming months or years, start budgeting for these now.

4. **Keep a Financial Buffer**:

    o   Maintain a buffer even after the initial contingency funds have been used. Continually adding to this buffer from any surplus in your budget can provide ongoing financial security.

Effective budget allocation, exploring funding options, contingency planning, and regular budget reviews are essential components of a successful transition to off-grid living. By carefully planning and managing your finances, you can ensure that your off-grid lifestyle is sustainable and resilient, allowing you to enjoy the benefits of self-sufficiency without financial stress. This disciplined approach to budgeting not only secures your off-grid journey but also enhances your ability to adapt and thrive in your new environment.

Choosing the Right Location

Choosing the right location is essential for the success and sustainability of an off-grid lifestyle. It significantly impacts the construction of your home, your daily life, and the long-term viability of living independently from municipal services. Here's a detailed exploration of the various considerations involved in selecting an optimal location for off-grid living.

Climate Suitability

Understanding the climate of potential locations is paramount as it directly affects your shelter, attire, and agricultural activities. For instance:

*   **Cold Climates**: In colder regions, you will need a well-insulated dwelling to maintain warmth. Heating solutions such as wood stoves or solar thermal systems are vital. Buildings might need to withstand heavy snowfall, requiring specific architectural styles like steep roofs.

- **Warm Climates**: Warmer areas might necessitate the implementation of cooling systems, such as passive solar designs that enhance airflow and shade from the sun. Materials that do not absorb much heat, like reflective roofing or light-colored exterior walls, can be beneficial.

- **Growing Seasons**: The length of the growing season is crucial if you plan to cultivate your own food. A longer growing season allows for a broader range of crop varieties and extended harvesting times, which can contribute to greater food security.

Natural Disaster Risks

The risk of natural disasters can significantly influence the choice of location:

- **Flood Zones**: Areas prone to flooding require elevated homes and special considerations in terms of insurance and emergency planning.

- **Earthquake Prone Areas**: If earthquakes are a risk, your building design will need to incorporate earthquake-resistant construction techniques.

- **Wildfire Risks**: In regions susceptible to wildfires, choosing a location away from dense forests and implementing a defensible space around your property can mitigate risks.

- **Severe Storms**: Places frequently hit by severe storms might need reinforced structures and storm shelters.

Water Access

Access to clean, reliable water is a cornerstone of off-grid living:

- **Proximity to Natural Water Sources**: Properties near streams, rivers, or lakes are ideal. These sources provide a steady supply of water but require proper filtration systems to ensure water quality.

- **Legalities of Water Use**: Before settling in, it's important to understand the legal aspects of water rights and usage in your intended area. Some regions have strict regulations on the use of natural water sources.

- **Rainfall Patterns**: Knowing the rainfall patterns helps in designing effective rainwater harvesting systems. Areas with higher annual rainfall can support larger collection systems, providing ample water for household and agricultural needs.

Land Characteristics and Resources

Choosing a suitable location for an off-grid lifestyle also means paying close attention to the land's specific characteristics and the resources it offers. Here's an in-depth look at the critical aspects of soil quality, availability of natural resources, and energy potential, each of which plays a crucial role in the sustainability and feasibility of off-grid living.

Soil Quality

The foundation of any agricultural activity is the soil. Its health determines the variety and abundance of crops you can grow, which directly impacts your food security.

- **Soil Testing**: Before committing to a location, conduct a comprehensive soil test to evaluate its fertility, pH level, and presence of contaminants. This will help you understand what kind of soil amendments or treatments are necessary.

- **Type of Soil**: Different soil types, from loamy to clay, offer distinct advantages and challenges. For example, loamy soil is ideal for most agricultural crops due to its optimal drainage and nutrient retention.

- **Amendments and Management**: Depending on the soil test results, you might need to consider the addition of organic matter, such as compost or manure, to improve soil fertility and structure.

Availability of Resources

The local ecosystem can provide much of what you need to live sustainably off the grid. Assessing the availability of these resources is essential for planning your home and lifestyle.

- **Timber and Firewood**: In wooded areas, access to timber can solve many construction and heating needs. However, sustainable management practices are crucial to prevent deforestation and to maintain ecological balance.

- **Stone and Clay**: Natural deposits of stone and clay can be valuable for building durable structures and for crafts such as pottery, which can add functional and aesthetic value to your living environment.

- **Water Sources**: Beyond drinking and household use, water bodies support agriculture and can also be a source of food (e.g., fish).

Energy Potential

The potential to harness renewable energy sources is a determining factor in the viability of living off the grid.

- **Solar Power**: Areas with high solar insolation are prime candidates for solar energy systems. The orientation of the land and lack of shade will affect the efficiency of solar panels.

- **Wind Energy**: Locations on higher ground or open plains may be more suitable for wind turbines, which require unobstructed wind flows to operate efficiently.

- **Hydroelectric Power**: If your land includes or is near a flowing river, micro-hydroelectric power can be a reliable and continuous source of energy.

Legal and Community Aspects

When planning for an off-grid life, understanding the legal requirements and the community dynamics of potential locations is crucial. This not only ensures compliance with local regulations but also facilitates a smoother integration into a new environment. Here's a detailed exploration of zoning, building codes, community relationships, and accessibility issues.

Zoning and Building Codes

The legal landscape of your chosen location can significantly influence your off-grid plans, especially concerning what you can build and how you can live sustainably.

- **Local Zoning Laws**: These laws determine the usage of land within specific areas. For instance, some zones might be restricted to agricultural use, while others may allow residential buildings but with strict limitations on the type and size of structures.

- **Building Codes**: Compliance with local building codes is mandatory and can dictate everything from structural requirements to the minimum standards for electrical and plumbing systems. These codes are designed to ensure safety and sustainability but can vary widely between different areas.

- **Off-Grid Specific Restrictions**: Some locales may have specific restrictions affecting off-grid living, such as bans on rainwater collection or limitations on the type of sewage systems you can install. These restrictions might be due to environmental concerns or local infrastructure capabilities.

Community and Accessibility

While seeking independence is a key aspect of off-grid living, the proximity to and relationship with nearby communities can have significant benefits, particularly in terms of security, emergency support, and social well-being.

- **Emergency and Medical Access**: In case of emergencies, being within a reasonable distance of medical facilities can be life-saving. Also, consider the response time for emergency services when choosing your location.

- **Social Interactions**: Human beings are inherently social; thus, access to a community can help mitigate feelings of isolation associated with remote living. It provides opportunities for socializing, sharing resources, and mutual support.

- **Accessibility and Transportation**: Evaluate the accessibility of the location throughout the year. Some areas might become cut off during certain seasons due to snow or flooding. Accessibility affects not just daily convenience but also the feasibility of emergency evacuations or receiving supplies.

- **Road Infrastructure**: The state of roads and availability of public transportation can significantly influence your quality of life. Well-maintained roads are crucial for the regular transport of supplies, accessing services, and integrating with the community.

Additional Considerations

- **Local Support Networks**: Some communities are more welcoming and supportive of off-grid lifestyles than others. Engaging with local groups or attending community meetings can provide insights into the community's dynamics and potential support systems.

- **Cultural Fit**: Every community has its own cultural norms and values, which can affect how well you integrate and feel accepted. Understanding these cultural aspects can help in making a more informed decision about where to settle.

Personal Preferences and Long-Term Goals

When contemplating a move to an off-grid lifestyle, personal preferences and long-term aspirations play a pivotal role in determining the perfect location. Here's a comprehensive look at how lifestyle compatibility, future development, and investment considerations can influence your decision.

Lifestyle Compatibility

Choosing a location that harmonizes with your personal lifestyle and future goals is essential for long-term satisfaction and well-being in an off-grid setting.

- **Outdoor Recreation**: If you're an enthusiast of outdoor activities like hiking, fishing, or kayaking, selecting a location near natural landscapes such as forests, mountains, or bodies of water can enrich your daily life. These environments not only offer recreational opportunities but also contribute to your mental and physical health.

- **Tranquility and Seclusion**: Consider the level of solitude or community interaction you desire. A secluded spot far from urban centers can offer peace and quiet, ideal for those looking to escape the hustle and bustle of city life. However, this may also mean fewer social interactions and longer trips for basic supplies or services.

- **Climate and Comfort**: The local climate can affect your daily comfort and the efficiency of your off-grid systems. For example, colder regions might require more heating and robust insulation strategies, whereas warmer areas might demand effective cooling and shade solutions to maintain comfort.

Future Development

Understanding the potential for future changes in the area can help you anticipate how your chosen location might evolve over time, affecting both the natural environment and your quality of life.

- **Infrastructure Developments**: Upcoming infrastructure projects like new roads or utilities can alter the landscape significantly, impacting the natural beauty and isolation of an area.

- **Industrial Expansion**: Be aware of any planned industrial activities or developments that could introduce pollution, noise, or increased traffic, which could detract from the peaceful off-grid lifestyle you seek.

- **Population Growth**: Areas on the brink of significant population increases might face challenges like increased land prices, reduced privacy, and greater competition for resources, which could affect your living conditions and self-sufficiency goals.

Resale Value and Investment

While the primary motivation for moving off-grid might be a sustainable and independent lifestyle, it's pragmatic to consider the financial aspects of your investment.

- **Market Trends**: Analyze the real estate trends in the area to gauge potential resale values. Properties in popular or up-and-coming areas might appreciate in value, offering a good return on investment if you ever decide to sell.

- **Land Desirability**: The features of the land itself, such as water access, scenic views, and overall usability, can significantly influence its marketability and appeal to future buyers.

- **Economic Stability**: Regions with stable or growing economies tend to maintain or increase property values over time, making them safer financial investments.

Choosing the right location is fundamental to the success and sustainability of an off-grid lifestyle. This decisive step influences nearly every aspect of your life, from the construction of your home to your daily activities, and the long-term viability of your independence from traditional municipal services. By carefully evaluating climate suitability, understanding the risks of natural disasters, ensuring access to essential resources like water, and taking into account land characteristics, you can select a location that not only meets your immediate needs but also supports your future aspirations. Additionally, considering legal constraints and community dynamics ensures that your chosen location will allow you to live in compliance with local regulations and integrate into the community as needed. Making a well-informed decision on location sets the foundation for a successful, sustainable off-grid living arrangement.

Book 2: Water

Methods for Rainwater Collection

Basic Setup and Components

Rainwater collection is a fundamental aspect of sustainable living, especially for those opting for an off-grid lifestyle. An effective rainwater collection system begins with a well-designed basic setup consisting of several key components: the catchment area, gutters and downspouts, and first flush diverters. Each component plays a pivotal role in ensuring that the collected rainwater is of good quality and suitable for various uses.

1. Catchment Area

The catchment area, typically the roof of a house, is the primary surface from which rainwater is collected. The efficiency and safety of rainwater collection heavily depend on the material and condition of the catchment area.

- **Materials**: Metal roofing is often preferred for rainwater collection due to its durability and relatively inert properties, which minimize the risk of leaching harmful chemicals. Treated tiles can also be used, although it's important to ensure that the treatment does not introduce substances that could contaminate the water. Other materials like slate or synthetic roofing can also be effective, provided they do not affect water quality.

- **Surface Area and Design**: The size of the catchment area directly influences the volume of water that can be collected. Larger roof areas will collect more water, thereby increasing the water supply. The design of the roof should also facilitate efficient water flow towards the gutters, avoiding areas where water might pool and become contaminated.

- **Maintenance**: Regular maintenance of the catchment area is crucial. This includes cleaning the roof to remove dust, leaves, and other debris, which could block the flow of water or degrade its quality. Inspecting the roof for damage and ensuring it is in good repair can prevent contaminants from entering the water supply.

2. Gutters and Downspouts

Gutters and downspouts are essential for channeling water from the catchment area to the storage tanks. Proper installation and maintenance of these components are key to maximizing water collection and preventing water loss.

- **Materials and Installation**: Gutters are typically made from materials such as PVC, aluminum, or stainless steel. They need to be securely attached to the roof's edge and sloped correctly to ensure that water flows freely towards the downspouts, which then direct the water to the storage system.

- **Capacity and Scalability**: The size and number of gutters and downspouts should be adequate to handle the maximum amount of water expected during heavy rainfall. This prevents overflow and ensures that all collectible water is directed to the storage tanks. In areas with heavy rainfall, larger or additional gutters might be necessary.

- **Maintenance**: Gutters should be regularly cleaned to remove leaves, bird nests, and other debris that could cause blockages. Downspouts should also be checked to ensure they are not obstructed and are directing water correctly. This maintenance is crucial to prevent water overflow and damage to the building structure.

3. First Flush Diverters

First flush diverters are a critical component of any rainwater collection system. They improve water quality by diverting the initial flow of water, which may contain higher levels of dust, bird droppings, and other contaminants, away from the storage tanks.

- **Function and Importance**: The first flush of water from the roof can carry a disproportionate amount of the accumulated contaminants. By diverting this first flush into a separate drainage or storage system, the overall quality of the collected water is significantly enhanced.

- **Design and Operation**: A typical first flush diverter includes a valve that automatically closes once a predetermined amount of water has been diverted. This amount is usually based on the surface area of the roof and the average buildup of contaminants between rain events.

- **Maintenance and Configuration**: Regular inspection and maintenance of first flush diverters are essential to ensure their effective operation. This includes cleaning out any debris that may have been diverted and checking that the valve mechanisms are functioning correctly.

Storage Solutions

Implementing efficient storage solutions for rainwater is a cornerstone of sustainable off-grid living. Whether you are setting up a small garden or planning for substantial agricultural use, having the right storage system can ensure an adequate and reliable water supply. Here's an in-depth look at two popular storage solutions: rain barrels and cisterns, including their advantages, materials, and maintenance requirements.

Rain Barrels: Simple and Economical Water Storage

Rain barrels are an accessible starting point for rainwater collection. Typically, these are large barrels placed directly beneath downspouts to catch rainwater flowing off the roof. Here's how to maximize their efficiency:

1. Setup and Installation:

- Choose a UV-resistant material like polyethylene to prevent algae growth and prolong the barrel's life.

- Position the barrel on a stable, elevated platform to facilitate gravity-fed water flow and easy access to the spigot.

- Ensure that the barrel has a screen cover to keep out debris and insects.

2. Capacity and Scalability:

- Standard rain barrels hold about 50 to 80 gallons, but linking multiple barrels can increase capacity.

- Consider using diverter kits to connect additional barrels, ensuring a larger reservoir and more consistent water pressure.

3. Maintenance and Usage:

- Regularly clean the gutels and downspouts to prevent clogging and contamination.

- Use the collected water for non-potable purposes like irrigation, washing cars, or watering lawns.

- During winter, empty the barrels to prevent damage from freezing.

4. Benefits:

- Rain barrels are cost-effective, often available at low costs, or you can repurpose food-grade containers.

- They reduce the runoff, which can decrease erosion and lessen the burden on stormwater systems.

Cisterns: Robust Solutions for Larger Scale Water Needs

For those requiring a more substantial water storage capacity, cisterns offer a versatile and durable solution. They can be installed above or below ground and are designed to hold anywhere from a few hundred to thousands of gallons of water.

1. Types of Cisterns:

- **Above Ground:** Easier to install and maintain, ideal for areas with stable, mild climates as they are susceptible to temperature extremes.

- **Below Ground:** Protected from the elements, making them suitable for freezing climates. They also save space and preserve the aesthetic of your landscape.

2. Materials:

- **Plastic/Fiberglass:** Lightweight, resistant to corrosion, and relatively inexpensive. They are quick to install but may require anchoring to prevent buoyancy issues in waterlogged soils.

- **Concrete:** Extremely durable and can be customized to any size. Concrete cisterns are often installed underground and can improve the pH balance of the stored water.

- **Metal:** Typically steel with a corrosion-resistant coating, metal cisterns are sturdy but can be expensive and prone to rust if the coating is damaged.

3. Installation Considerations:

- Professional installation is recommended, especially for large or underground cisterns to ensure proper sealing and longevity.

- Incorporate overflow mechanisms and filters to manage excess water and maintain water quality.

4. Maintenance:

- Inspect regularly for leaks, especially at seams and connections.

- Clean the interior periodically to remove sediment and prevent algae growth.

- Check and clean filters and screens to ensure efficient operation.

5. Benefits:

- Provides a significant volume of stored water, ensuring availability during dry periods.

- Enhances property value with a permanent solution for water conservation.

Integrating Rain Barrels and Cisterns

To create a comprehensive rainwater harvesting system, consider integrating both rain barrels and cisterns. Start with rain barrels for immediate, small-scale collection and usage. As your needs and capacity grow, expand your system with a cistern, providing a more substantial and stable water supply. This approach allows for flexibility and scalability, adjusting to your water needs as they evolve over time.

Maintenance and Safety

Maintaining a rainwater collection system is crucial for ensuring the safety and quality of the stored water and for prolonging the system's lifespan. This maintenance includes a series of regular cleaning routines, as well as measures to control pests and algae growth. Here's a comprehensive look at the processes and safety measures involved in maintaining a rainwater collection system.

Regular Cleaning

Consistent maintenance routines are essential to keep rainwater collection systems functional and the water they store clean and safe for use.

1. Catchment Area and Gutters:

- **Debris Removal:** Regularly remove leaves, twigs, and other debris from the roof and gutter system. This prevents blockages and reduces the organic load entering the storage tanks, which can decompose and contaminate the water.

- **Gutter Guards:** Installing gutter guards can help reduce the amount of debris that collects in the gutters, thereby minimizing maintenance frequency and improving water quality.

- **System Inspection:** Periodically inspect the catchment area and gutters for damage or wear. Look for cracks, holes, or any signs of rust and repair them promptly to prevent contaminants from entering the system.

2. First Flush Diverters:

- **Regular Checks:** Ensure that the first flush diverters are functioning correctly. These devices should be checked and cleaned out after each significant rainfall to ensure they effectively remove the dirtiest initial water.

- **Maintenance Schedule:** Clean the first flush mechanism at least twice a year or more frequently in areas with high dust or frequent bird activity.

3. Storage Tanks:

- **Tank Cleaning:** Over time, sediment can build up at the bottom of storage tanks. This sediment should be removed periodically to prevent anaerobic processes and the proliferation of bacteria. Professional tank cleaning services are recommended, especially for larger tanks.

- **Inspection for Leaks:** Check tanks regularly for any signs of leakage or corrosion, particularly in metal tanks, which are prone to rust.

Mosquito and Algae Control

Preventing the growth of algae and the breeding of mosquitoes in rainwater storage tanks is critical for maintaining water quality and ensuring the safety of the system.

1. Mosquito Control:

- **Secure Lids:** All tank openings should be covered with tight-fitting lids to prevent mosquitoes from entering and breeding in the water.

- **Screens:** Install fine mesh screens on any vents or overflows. This will allow air to circulate while keeping mosquitoes out.

- **Biological Control:** Introducing mosquito larvae-eating fish (like gambusia) or using bacterial larvicides (such as products containing *Bacillus thuringiensis israelensis*) can effectively control mosquito populations without harming the water quality.

2. Algae Control:

- **Tank Material and Placement:** Use opaque materials for storage tanks and place them in areas with minimal direct sunlight to reduce algae growth.

- **Regular Cleaning:** Keep the interior and exterior of tanks clean and free from organic matter, as this can serve as a nutrient source for algae.

- **Algaecides:** If algae growth becomes a problem, algaecides can be used, but it is crucial to choose products that are safe for the intended use of the water. For instance, copper sulfate can be effective against algae but must be used cautiously, especially if water is to be used for potable purposes.

Safety Measures

Implementing safety measures is essential to protect both the system and the health of its users.

1. Chemical Storage and Use:

- Ensure that any chemicals used in the system, such as algaecides or cleaning agents, are stored safely away from children and pets.

- Use only chemicals approved for use in potable water systems if the collected rainwater is intended for drinking.

2. Regular Testing:

- Regularly test the water quality for pathogens, chemicals, and heavy metals, especially if it is used for drinking or cooking. This will help identify any potential issues early.

3. Training and Awareness:

- Educate all household members about the proper operation and maintenance of the rainwater collection system.

- Maintain an operations manual and a schedule for regular maintenance activities.

Effective maintenance and safety protocols are essential for the successful operation of a rainwater collection system. Regularly scheduled cleaning, diligent mosquito and algae control, and strict safety measures ensure that the water remains clean, safe, and suitable for use, thereby maximizing the benefits of raincare collection.

System Integration and Backup

System integration and backup planning are crucial aspects of a robust rainwater collection system, particularly for those seeking a resilient, sustainable approach to water management. Combining rainwater systems with other water sources and establishing effective backup strategies can greatly enhance the reliability and efficiency of water supply in both residential and agricultural settings. Here's a detailed exploration of how to integrate rainwater collection systems with other solutions and the importance of having a backup plan.

Integration with Other Systems

Integrating rainwater collection with other water systems involves creating a hybrid setup that can optimize water usage and ensure a consistent supply even under varying environmental conditions.

1. Combination with Well Water:

- **Complementary Use:** In areas where groundwater is accessible, combining rainwater with well water can balance the reliance on each source. For instance, rainwater can be primarily used for non-potable applications such as irrigation and washing, while well water can be reserved for drinking and cooking.

- **System Design:** The integration requires a well-designed plumbing system with valves and switches that can easily switch between the two sources or mix them as needed.

2. Greywater Recycling:

- **Water Reuse:** Greywater from baths, sinks, and washing machines can be treated and reused for garden irrigation and toilet flushing. Integrating this with a rainwater collection system can maximize water efficiency.

- **Treatment and Storage:** It's essential to have proper filtration and disinfection systems in place to ensure that recycled greywater meets safety standards for its intended use.

3. Municipal Water as a Supplement:

- **Emergency Use:** In urban settings where municipal water is available, it can serve as a backup to rainwater during periods of low rainfall.

- **Regulatory Compliance:** Ensure that the system complies with local water use regulations, particularly concerning the connection between private and public water systems.

4. Smart System Integration:

- **Technology Use:** Employing smart technologies like automated valves, water quality sensors, and IoT devices can help in efficiently managing water resources from multiple sources, optimizing usage based on availability, demand, and quality.

Backup Plans

Having a well-thought-out backup plan is essential for ensuring water availability during emergencies, such as prolonged droughts or mechanical failures in the water collection system.

1. Additional Water Reserves:

- **Larger Storage Capacity:** Consider installing additional water storage tanks to increase your water holding capacity. This can be a safeguard against periods of low rainfall.

- **Underground Cisterns:** For long-term storage, especially in areas prone to evaporation, underground cisterns can be beneficial as they minimize water loss and maintain lower temperatures to inhibit algae growth.

2. Alternative Water Sources:

- **Portable Water Deliveries:** In extremely dry regions or during unexpected system failures, arranging for water delivery can be a viable backup option.

- **Natural Water Bodies:** If located near rivers, lakes, or streams, developing a method for drawing water from these sources during emergencies can be part of a comprehensive backup strategy. Legal permissions and water treatment before use should be considered.

3. System Redundancy:

- **Duplicate Critical Components:** Installing duplicate systems for key components such as pumps and filters can prevent complete system shutdowns due to mechanical failure.

- **Manual Overrides:** Ensure that systems can be operated manually in case of power failures or electronic control malfunctions.

4. Rainfall Capture Expansion:

- **Additional Catchment Areas:** Expanding the catchment area by including additional structures like sheds or carports can increase water collection capacity.

- **Temporary Solutions:** Portable rain barrels and temporary catchment solutions can be deployed quickly in anticipation of forecasted heavy rainfall.

5. Emergency Preparedness:

- **Regular Testing and Drills:** Conduct regular tests of the backup systems and practice emergency procedures to ensure they are effective and familiar to all users.

- **Community Resources:** Engage with local community groups or water management authorities to understand additional resources or support available during water shortages.

Compliance with local regulations is a crucial step in setting up a rainwater collection system. It's essential to understand and adhere to any local laws and regulations regarding water collection and usage. Some regions might have specific restrictions or require permits for setting up such systems. Ensuring legal compliance not only avoids potential fines and legal issues but also helps in maintaining sustainable and environmentally friendly practices. By following these guidelines and maintaining your system effectively, you can secure a reliable and sustainable water source that supports your off-grid lifestyle, minimizes your environmental impact, and provides a dependable supply of water.

Water Filtration and Purification

Water filtration and purification are crucial in ensuring the safety and usability of water, especially for those living off-grid, where access to municipally treated water is unavailable. The essential nature of these processes lies in their ability to transform potentially hazardous water into a clean, consumable resource. This guide delves into various methods of filtration and purification tailored to different needs and environments, emphasizing the importance of these processes for off-grid living.

Understanding the Need for Filmentation and Purification

The main objective of water filtration and purification is to eliminate contaminants that may pose health risks or negatively impact the water's taste, odor, and appearance. These contaminants fall into three broad categories:

1. **Physical contaminants:** These include substances like soil, sediment, and organic material which can make water appear cloudy and unappealing. Physical filtration methods target these particulates, removing them from the water to enhance clarity and safety.

2. **Chemical contaminants:** Both naturally occurring and man-made chemicals such as nitrates, fluorides, pesticides, and heavy metals can be found in water sources. These chemicals can have detrimental effects on health, making their removal or reduction critical. Filtration methods such as activated carbon filters are particularly effective in reducing chemical contaminants by adsorbing them onto their surface.

3. **Biological contaminants:** Pathogens like bacteria, viruses, and protozoa are perhaps the most concerning as they can directly cause diseases. Purification methods are necessary to address these biological threats, ensuring the water is safe for consumption.

Filtration Techniques

Filtration techniques play a pivotal role in ensuring the quality and safety of water, especially in contexts where access to treated water is limited. Filtration not only clarifies water but also improves its taste and removes harmful contaminants. Below, we delve into various filtration methods that cater to different needs, explaining how each technique works and its specific applications. This comprehensive exploration helps in understanding the integral role of these systems in water treatment processes.

Sediment Filfiltration

Sediment filters are the first line of defense in water filtration systems. They primarily remove large particles such as dirt, rust, and sand from water. These filters are typically made from fibrous or granular materials which act as a sieve to trap solid particles. The effectiveness of sediment filtration largely depends on the micron rating of the filter, which indicates the size of particles it can effectively capture. Sediment filters are crucial for protecting more sensitive downstream filters from clogging.

Activated Carbon Filters

Activated carbon filters are renowned for their ability to improve water taste, remove odors, and reduce the presence of many chemical contaminants. These filters utilize a process known as adsorption, where contaminant molecules in the water are trapped on the surface of activated carbon. This form of filtration is particularly effective against chlorine, volatile organic compounds (VOCs), and certain pesticides and herbicides. Due to their porous nature, activated carbon filters also assist in reducing pollutants like lead and mercury when enhanced with specific additives.

Ceramiciffilters

Ceramic filters boast extremely small pore sizes, making them suitable for removing microbial contaminants such as bacteria, protozoa, and cysts from water. These filters often contain incorporated silver, which acts as a bacteriostatic agent to prevent the growth of microbes within the filter itself. Ceramic filters are durable, long-lasting, and can be cleaned and reused multiple times. They are particularly useful in remote areas where other means of purification may not be readily available.

Reverse Osmosis (RO)

Reverse osmosis represents a more sophisticated filtration technique that employs a semipermeable membrane to remove a wide array of contaminants, including dissolved salts, viruses, and bacteria. Water is forced under pressure through this membrane, leaving contaminants behind. The result is highly purified water. However, reverse osmosis systems require significant maintenance, including regular cleaning of the membrane and replacement of pre- and post-filters. Additionally, they are relatively water-intensive, producing a substantial amount of wastewater.

Membrane Frequencyration

Beyond traditional methods, there are innovative technologies such as ultrafiltration and nanofiltration that utilize membranes with different pore sizes. These are particularly adept at targeting specific contaminants—

ultrafiltration is excellent for removing proteins, viruses, and larger bacteria, whereas nanofiltration can remove smaller molecules like certain organic compounds and softening hard water.

Choosing the Right Filfitration Technique

Selecting the appropriate filtration system depends on several factors including the nature of the water source, the specific contaminants present, and the desired quality of the finished water. For most residential and off-grid applications, a combination of these filtration techniques can be deployed to achieve the best results. For instance, a system might integrate sediment and activated carbon filters with a reverse osmosis unit to provide comprehensive water treatment.

Purification Methods

Purification methods are essential components in ensuring the safety of drinking water, particularly in areas where access to municipally treated water is absent. This detailed guide covers a range of common purification techniques, each suited to different environments and needs. Understanding these methods is crucial for anyone relying on natural water sources, aiming to make their water consumption safe and sustainable.

Boiling

Boiling is one of the oldest and most reliable methods of water purification. It involves heating water to a rolling boil, typically for at least one minute at sea level and three minutes at altitudes above 6,500 feet. This process is effective because it kills or inactivates viruses, bacteria, protozoa, and other pathogens by denaturing their proteins and destroying their cell membranes. Boiling is universally accessible and does not require any special equipment beyond a heat source and a pot, making it an ideal method for emergency situations or daily use in remote areas.

Chemical Disinfection

Chemical disinfection involves the addition of chemicals such as chlorine or iodine to water to kill microbial pathogens. This method is effective against a wide range of microorganisms and is commonly used both in personal and large-scale water treatment systems.

- **Chlorine**: Chlorine tablets, liquid bleach, or granules are widely used due to their effectiveness, availability, and cost-efficiency. Chlorine not only kills bacteria and viruses but also some types of protozoa. The correct dosage and contact time are critical for effective disinfection, and there might be a residual taste that some find unpleasant.

- **Iodine**: Similar to chlorine, iodine is available in tablets, drops, and solutions. It is effective against viruses, bacteria, and many protozoa. Iodine is typically used in smaller, personal water purification efforts, such as for hikers and travelers. However, it imparts a distinctive taste to the water and is not suitable for pregnant women or individuals with thyroid problems.

Ultraviolet (UV) Light

UV purification systems use ultraviolet light to deactivate the DNA of bacteria, viruses, and other pathogens, preventing them from multiplying and causing disease. This method is highly effective, leaves no taste, and does not introduce chemicals into the water. However, it requires a power source, which can be a limitation in off-grid or remote locations. UV systems also require relatively clear water, as turbidity can shield microorganisms from UV light.

Solar Disinfection (SODIS)

Solar disinfection, or SODIS, is a simple, low-cost method that uses the sun's UV rays to kill pathogens in water. It involves filling clear plastic or glass bottles with water and exposing them to full sunlight for six hours (or two days if the sky is overcast). This method is particularly useful in tropical and subtropical regions where sunlight is abundant. SODIS is effective against various pathogens, including bacteria and viruses. However, it does require clear skies and intense sunlight, limiting its use in cloudy or less sunny climates.

Additional Methods

Beyond these common techniques, there are other methods of purification worth mentioning:

- **Distillation**: Distillation involves boiling water and then condensing the steam back into a liquid. This method removes impurities, including minerals, bacteria, viruses, and chemicals. Distillation units can be homemade or purchased and are particularly useful in removing salt from seawater or brackish water.

- **Advanced Filtration Systems**: Some advanced filters, like those incorporating nano-filtration technology, can remove not only particles but also pathogens and some chemical impurities. These systems often combine multiple stages of filtration and can even include UV light for a comprehensive treatment solution.

In conclusion, a variety of water purification methods are available, each with its advantages and limitations. The choice of a suitable purification technique depends on the specific contaminants present, the intended use of the water, and the resources available. Whether for daily use or emergency preparedness, understanding these methods is essential for ensuring the safety and palatability of your water supply.

Implementing a Multi-Barrier Approach

Implementing a multi-barrier approach to water treatment ensures the highest safety and quality of water, particularly crucial in environments where access to municipally treated water is limited. This strategy employs multiple filtration and purification techniques in tandem to effectively remove or neutralize a broad spectrum of contaminants, from physical sediments to biological pathogens. Here is a detailed exploration of how to effectively apply a multi-barrier approach, along with the necessary maintenance and monitoring practices to keep water systems functioning optimally.

Understanding the Multi-Barrier Approach

The multi-barrier approach to water treatment is designed to prevent the failure of any single treatment step from compromising the safety of the water supply. This method increases reliability through redundancy, ensuring that even if one barrier fails, additional barriers will provide protection. Typical configurations might include:

- **Sediment Filtration**: This first stage removes large particles such as dirt, rust, and sand, preventing them from affecting subsequent filters.

- **Activated Carbon Filtration**: Following sediment filtration, activated carbon filters out chemicals, odors, and tastes. It can also reduce certain organic compounds and chlorine.

- **Reverse Osmosis (RO)**: For a more thorough filtration, RO systems push water through a semi-permanent membrane that captures extremely small contaminants, including some bacteria and viruses.

- **UV Purification**: As a final barrier, UV light treatment effectively neutralizes all remaining microbiological contaminants, ensuring that the water is safe for consumption.

By integrating these systems, a multi-barrier approach addresses different contaminant types and ensures comprehensive water treatment.

Maintenance of Filtration and Purification Systems

To keep the multi-barrier systems operating efficiently, regular maintenance is essential. Here's what it typically involves:

- **Replacing Filters**: Each filter has a lifespan after which it must be replaced to ensure optimal functioning. Manufacturers usually provide guidelines on when to replace filters, but signs like reduced water flow or an unusual taste in the water can also indicate it's time for a change.

- **Cleaning Systems**: Mechanical parts such as pumps and valves, as well as storage tanks, need regular cleaning to prevent the buildup of sediments and biofilms which can harbor contaminants. This also includes inspecting and cleaning gutter systems and pre-filtration units that collect debris before it enters the storage tanks.

- **Inspecting UV Lamps**: UV lamps degrade over time and must be checked regularly to ensure they are producing sufficient UV light to inactivate pathogens effectively.

Monitoring Water Quality

Alongside regular maintenance, continuous monitoring of water quality is crucial. This ensures that all components are functioning properly and that the water remains safe to drink. Monitoring might involve:

- **Visual Inspections**: Regular checks for clarity, color, and sediment in water can indicate problems with the filtration stages.

- **Taste and Odor Tests**: Changes in the taste or odor of water can be early indicators of system failures or contamination.

- **Chemical Tests**: Using test kits to measure levels of specific contaminants like chlorine, pH, heavy metals, and nitrates can help assess the chemical safety of water.

- **Microbiological Tests**: Periodic testing for pathogens such as bacteria, viruses, and protozoa is essential to confirm the biological safety of the water.

In conclusion, water filtration and purification are essential for ensuring the safety and quality of water, particularly in off-grid scenarios where traditional municipal water systems are not accessible. The importance of these processes cannot be overstated, as they transform potentially hazardous water sources into safe, drinkable water. This comprehensive guide underscores the variety of methods available to filter and purify water, each suited to handling different types of contaminants—from physical debris to biological threats. By applying the techniques discussed, individuals can effectively manage their water needs in various environments, ensuring that their water supply is both safe and sustainable. Adopting such measures is not just a matter of convenience but a crucial aspect of health and well-being in off-grid living.

Irrigation Systems

Irrigation systems are vital components in the management of water resources, particularly in agriculture, gardening, and maintaining landscapes. Efficient irrigation is crucial for maximizing water usage, enhancing plant growth, and conserving water, especially in regions prone to drought or where water resources are scarce. This section explores various irrigation systems, each tailored to different needs and environmental conditions, and provides a guide on how to implement and manage these systems effectively.

Types of Irrigation Systems:

1. Drip Irrigation: Drip irrigation stands out as one of the most effective and efficient irrigation methods, especially suited for row crops, perennial plantations, and vegetable gardens. This targeted irrigation approach delivers water directly to the root zone of each plant through a sophisticated network of valves, pipes, tubing, and emitters, ensuring optimal water usage and promoting robust plant health.

Principles of Drip Irrigation

The essence of drip irrigation is its ability to place water directly where it's most needed—at the roots of the plants. This precision minimizes waste, reduces evaporation, and limits runoff, which are common issues with more traditional methods of irrigation like overhead sprinkling. By focusing the water directly on the soil above the root system, drip irrigation ensures that plants receive the water they need to grow without saturating the surrounding soil, which can lead to weed growth and soil erosion.

Components of a Drip Irrigation System

A typical drip irrigation setup includes several key components:

- **Water Source:** This can be any reliable water supply, but it must be filtered to ensure that the water is free of sediments that could clog the drip system.

- **Pump and Pressure Regulators:** These control the water pressure within the system to ensure it is at an optimal level for drip irrigation, preventing damage to the system and ensuring even water distribution.

- **Valves:** Used for turning the water flow on and off and can be automated or manually operated.

- **Filters:** Essential for removing particles from the water that could potentially clog the emitters.

- **Drip Tubing or Hose:** Designed to be flexible and durable; it is the main conduit that delivers water from the source to the emitters.

- **Emitters:** These are the endpoints of the drip irrigation system, designed to release water in a slow and steady drip directly to the plant's root zone. They come in various discharge rates depending on the water requirements of specific plants.

Advantages of Drip Irrigation

Drip irrigation offers numerous benefits, making it highly suitable for a wide range of agricultural activities:

- **Water Efficiency:** By delivering water directly to the root zone, it significantly reduces wastage, making it an environmentally friendly option.

- **Cost-effective:** Reduces water usage and thus lowers the cost of water bills; it also requires less labor than traditional watering methods once it's set up.

- **Improved Plant Health and Yield:** Direct water delivery ensures that plants receive the optimal amount of water, which improves their health and can lead to increased yields.

- **Weed Reduction:** Water is not distributed where weeds might grow, thus reducing their proliferation.

- **Flexibility and Scalability:** Systems can be easily expanded or modified as plants grow or as more plants are added to the garden.

Installation and Maintenance

Installing a drip irrigation system involves planning the layout to ensure every plant will be sufficiently watered. This includes determining the placement of main and sub-main pipes and the spacing of the emitters. Maintenance of a drip irrigation system involves regular checking for leaks, flushing the system to remove any debris, and cleaning or replacing filters and emitters as needed to keep the system running efficiently.

2. Sprinkler Systems: Sprinkler systems, a common and versatile method of irrigation, effectively mimic natural rainfall by dispersing water through a network of pipes and spray heads. This method is especially suited for irrigating large areas like lawns, sports fields, and extensive agricultural lands. Here's an in-depth look at sprinkler irrigation, covering its mechanics, benefits, varieties, and best practices.

How Sprinkler Systems Work

Sprinkler irrigation works by pumping water under pressure through a system of pipes, leading to spray heads that distribute the water across the soil surface. The water is sprayed into the air, breaking into small droplets that fall to the ground, simulating the effect of rain. This method allows for uniform water distribution over a wide area, making it particularly effective for grass and other groundcovers which benefit from direct overhead watering.

Components of Sprinkler Systems

A typical sprinkler system includes the following key components:

- **Water Source:** This can be from municipal supplies, wells, ponds, or rainwater collection systems.

- **Pump Unit:** Responsible for providing the necessary pressure to drive water through the system.

- **Pipes and Valves:** Pipes transport water to different areas, and valves control the flow, allowing parts of the system to be shut off when not in use.

- **Sprinkler Heads:** These devices are strategically placed to ensure even coverage. They come in various types and sizes, designed to handle different flow rates and spray patterns.

- **Controller:** In automated systems, the controller is used to set the timing and duration of irrigation. Advanced systems can adjust schedules based on weather conditions or soil moisture levels.

Advantages of Sprinkler Systems

Sprinkler systems offer several significant advantages:

- **Efficiency:** They can cover large areas quickly and uniformly, making them highly efficient for watering large plots of land.

- **Versatility:** Suitable for a variety of landscapes, including residential lawns, agricultural fields, and commercial properties.

- **Customizability:** Systems can be customized with different types of sprinkler heads and programmable controls to meet specific watering needs.

- **Dust Control and Cooling:** Beyond irrigation, sprinklers can be used to control dust in dry areas and help cool down environments during hot weather.

Types of Sprinkler Systems

Sprinkler systems range from simple portable units to sophisticated automated and fixed installations:

- **Portable Sprinklers:** These are easy to move around and are ideal for small or irregularly shaped areas.

- **Fixed Sprinklers:** Installed permanently in the ground, these are typically used for residential lawns and commercial landscapes.

- **Traveling Sprinklers:** These systems move through the area to be irrigated, ideal for larger fields where permanent installations would be impractical.

- **In-Ground Sprinkler Systems:** These are installed below the surface, with pop-up sprinkler heads that only appear when the system is active.

Maintenance and Best Practices

To ensure efficiency and longevity of a sprinkler system, regular maintenance is crucial:

- **Regular Inspections:** Check for leaks, broken sprinkler heads, or any signs of malfunction.

- **Cleaning and Adjustments:** Clean filters and adjust spray heads to ensure even coverage without overspray onto buildings, pavements, or non-targeted areas.

- **Winterization:** In colder climates, systems should be drained and blown out with compressed air to prevent freezing and cracking in the pipes.

3. Surface Irrigation: Surface irrigation is a traditional and widely used agricultural practice where water is distributed over the land through the force of gravity. It's an especially prevalent method in areas that have ample water supplies and large, flat fields. This method is most commonly seen in the irrigation of crops like rice, which thrive in water-abundant environments. Here's a detailed look at surface irrigation, including its various types, how it works, its benefits, and the challenges it poses.

Basics of Surface Irritation

In surface irrigation, water flows over the soil surface to the points of infiltration. The efficiency of this method hinges on the field's ability to evenly distribute water across its surface, making it crucial for the land to be level. Here are the fundamental components that typically make up a surface irrigation system:

- **Water Source:** Can be rivers, reservoirs, or canals.

- **Distribution System:** Consists of channels or pipes that deliver water from the source to the fields.

- **Field Application:** The water is applied directly to the soil surface using various methods depending on the landscape and crop requirements.

Types of Surface Irrigation

There are three main types of surface irrigation, each suited to different environments and agricultural needs:

1. **Furrow Irrigation:**

   - **Description:** Water is directed into furrows or small channels dug between crop rows.

   - **Advantages:** Reduces evaporation by limiting the exposed water surface and is more efficient for row crops.

   - **Usage:** Commonly used for crops like maize, soybean, and cotton.

2. **Flood Irrigation:**

   - **Description:** Water covers the surface of the field to a certain depth.

   - **Advantages:** Simple to construct and operate; ideal for crops requiring a lot of water, like rice.

   - **Challenges:** Requires a lot of water, can lead to water wastage through runoff and deep percolation.

3. **Basin Irrigation:**

   - **Description:** Involves constructing small basins, each holding water around a group of plants or a single tree.

   - **Advantages:** Particularly effective for fruit trees and plantations.

   - **Control:** Water levels can be carefully controlled, minimizing the potential for water stress in the plants.

Benefits of Surface Irrigation

- **Low Cost:** Requires less infrastructure and technology compared to more modern irrigation methods like drip or sprinkler systems.

- **Simplicity:** Easy to manage and maintain with minimal technical skills.

- **Wide Applicability:** Can be used for a variety of crops and is especially advantageous in regions with abundant water.

Challenges and Considerations

Despite its benefits, surface irrigation has several drawbacks:

- **Water Efficiency:** It is generally less water-efficient, with higher evaporation rates and potential for runoff, especially if not carefully managed.

- **Land Suitability:** Requires relatively flat terrain to prevent water pooling and uneven distribution, which can lead to both waterlogging and under-watering.

- **Soil Erosion:** Intensive water flow can lead to soil erosion, particularly if the field isn't properly leveled.

- **Water Quality:** Poor water management can lead to salinization and alkalization of soils, particularly in arid regions.

Enhancing Efficiency

To increase the efficiency of surface irrigation systems, several strategies can be employed:

- **Levelling Fields:** Ensures even distribution of water, reducing wastage and improving infiltration.

- **Improved Water Control Structures:** Devices such as gates and valves can help regulate flow rates and volumes.

- **Scheduled Irrigation:** Aligning irrigation schedules with crop water needs and evaporation patterns can reduce water loss

4. Subsurface Irrigation: Subsurface irrigation is an advanced agricultural watering technique where water is delivered directly to the plant roots below the soil surface. This method uses a network of porous pipes or tubes placed beneath the soil to efficiently supply water to the root zone, significantly reducing evaporation losses and maximizing water usage efficiency. Here's a comprehensive overview of subsurface irrigation, including its mechanisms, advantages, applications, and considerations.

Mechanism of Subsurface Irrigation

In subsurface irrigation, a network of perforated pipes or drip tubes is buried at a strategic depth within the root zone of crops. These installations allow water to seep directly into the soil, targeting the root systems where it is most needed. The system can be designed to function through capillary action or with the aid of pressure, ensuring that water is distributed evenly and according to the specific moisture requirements of different crops.

Key Components

- **Water Source:** Includes well, reservoir, or municipal supply.

- **Pumping System:** Delivers water to the irrigation system, often equipped with filters to prevent clogging.

- **Distribution Network:** Consists of drip lines or porous pipes, buried at appropriate depths to target specific root zones.

- **Control Valves and Sensors:** Manage the flow and pressure within the system, often automated to respond to soil moisture levels.

Advantages of Subsurface Irrigation

1. **Water Efficiency:** By delivering water directly to the root zone and reducing surface evaporation, subsurface irrigation is among the most water-efficient methods available.

2. **Reduced Weed Growth:** Limiting surface water discourages the development of weeds, reducing competition for resources and minimizing the need for herbicides.

3. **Prevention of Soil Erosion:** With water applied below the surface, there is minimal water runoff, thereby preventing soil erosion and nutrient leaching.

4. **Water Conservation:** This system is ideal for regions with limited water resources or those prone to drought, as it maximizes the use of available water.

5. **Flexibility:** Subsurface irrigation can be used in various soil types and topographic conditions. It is especially beneficial for sloped terrain where surface methods might lead to uneven water distribution.

Ideal Applications

- **High-Value Crops:** Suitable for orchards, vineyards, and high-value vegetable crops, where precise water management is crucial.

- **Landscapes and Turf:** Commonly used in residential and commercial landscaping, including golf courses and public parks.

- **Areas with High Evaporation Rates:** Particularly effective in arid and semi-arid regions where evaporation rates are high.

Installation and Maintenance Considerations

While subsurface irrigation offers numerous benefits, it requires careful planning, installation, and ongoing maintenance to operate effectively:

- **Installation:** Proper installation is critical and often requires professional assistance to ensure that the system is correctly configured for the specific soil characteristics and crop types.

- **Maintenance:** Regular checks and maintenance are necessary to prevent clogging of the porous elements and to repair any leaks or breaks in the system. This includes monitoring for root intrusion and sediment buildup.

- **Cost:** The initial setup cost can be higher than surface irrigation systems due to the complexity of the components and the need for precise installation.

5. Micro Spray and Micro Sprinkler Systems: Micro spray and micro sprinkler systems represent an innovative solution in irrigation technology, tailored to meet the precise watering needs of smaller, more localized areas such as nurseries, greenhouses, and orchards. These systems combine the functionality of traditional sprinkler systems with enhanced control and efficiency, making them ideal for applications where delicate plants require gentle and precise watering. Here's a comprehensive overview of micro spray and micro sprinkler systems, including their design, benefits, applications, and considerations for use.

Design and Operation

Micro spray and micro sprinkler systems are designed to deliver water in a fine, controlled spray, allowing for targeted irrigation that minimizes waste and maximizes water usage efficiency. The system typically comprises:

- **Micro Sprayers and Sprinklers:** These are smaller than standard sprinklers and can be adjusted to control the radius and pattern of the water spray.

- **Tubing:** Flexible tubing delivers water directly to each micro sprayer or sprinkler, which is typically connected via a network that can be easily customized to suit specific layout needs.

- **Valves and Controllers:** Automated valves and precision controllers manage the flow of water, enabling the system to respond to the specific hydration needs of the plants, often with timers or moisture sensors that dictate when and how much to water.

Benefits of Micro Spray and Micro Sprinkler Systems

1. **Precision Watering:** By providing a controlled spray directly to the root zone of plants, these systems ensure that water is used efficiently, reducing runoff and evaporation.

2. **Adjustability:** Operators can adjust the flow rate and spray pattern to accommodate specific plant needs, which is especially beneficial in diverse plant settings like nurseries.

3. **Reduced Water Usage:** These systems are highly efficient, using less water than traditional sprinkler systems by focusing on targeted areas rather than indiscriminate coverage.

4. **Increased Humidity Control:** In environments like greenhouses, where humidity levels are crucial, micro sprays can maintain optimal moisture levels in the air, benefiting plant health and growth.

5. **Gentle on Plants:** The soft spray is ideal for young, fragile, or sensitive plants that might be damaged by the more intense water pressure of standard sprinkler systems.

Applications

- **Nurseries:** Provide gentle watering for young plants, which are particularly vulnerable to over-watering and water stress.

- **Greenhouses:** Maintain high humidity levels and precise moisture control to optimize plant growth conditions.

- **Orchards:** Target specific tree root systems without wasting water on non-productive areas.

- **Flower Gardens:** Adjust to the needs of different types of flowers, accommodating varying water requirements.

Installation and Maintenance

Installing micro spray and micro sprinkler systems requires careful planning to ensure that the layout matches the specific needs of the area and the types of plants being irrigated. Considerations include:

- **System Design:** Design should be based on the topography, plant type, and size of the area to ensure uniform water distribution.

- **Water Pressure:** Adequate pressure is necessary to ensure that each micro sprayer or sprinkler functions correctly, which may require the installation of pressure regulators.

- **Regular Maintenance:** Systems should be checked regularly for any clogs, leaks, or malfunctions, particularly since the small nozzles of micro sprayers are prone to blockage from debris or mineral buildup.

Implementing and managing an effective irrigation system is a multifaceted process that involves thorough planning and attention to several crucial factors. Efficient irrigation not only supports sustainable water use but also enhances plant health, which is vital in both agricultural and landscaping settings. Here's a detailed guide on how to approach the implementation and management of an irrigation system effectively.

Understanding Soil Type and Water Requirements

Before installing any irrigation system, it is essential to understand the type of soil and the specific water needs of the plants. Soil types vary greatly in their water absorption and retention capacities:

- **Sandy soils** drain quickly and require frequent but small watering sessions.

- **Clay soils**, on the other hand, retain water longer and need less frequent watering.

- **Loamy soils** offer a balance between water retention and drainage, which can affect the timing and amount of irrigation.

Additionally, different plants have varying requirements for water based on their species, life stage, and health. This knowledge will help in designing an irrigation system that matches the precise needs of the landscape or crop, thus optimizing water usage and promoting healthier plant growth.

Climate and Weather Patterns

The local climate and weather patterns play a significant role in irrigation management. For areas prone to drought, systems that conserve water, such as drip irrigation, are preferable. In contrast, regions with abundant rainfall might require systems that are adaptable to less frequent use. It's also important to consider seasonal weather variations and integrate systems that can adjust to changing conditions, such as rain sensors that reduce watering in rainy weather.

System Design and Layout

Proper design and layout of an irrigation system are critical to ensure uniform and efficient water distribution. A well-planned system should:

- **Cover all planted areas** without overspray onto sidewalks, roads, or buildings.

- **Be scalable** to accommodate future landscape changes or plant growth.

- **Include zoning** based on plant type and water needs, ensuring that each zone can be controlled independently to prevent overwatering or underwatering.

Maintenance of the Irrigation System

Regular maintenance is key to the longevity and efficiency of an irrigation system. Routine checks and upkeep should include:

- **Inspecting and cleaning filters** to prevent clogging and ensure smooth water flow.

- **Checking for leaks** in pipes and fittings, which can lead to significant water waste.

- **Adjusting sprinkler heads** to ensure they are providing adequate coverage and are not blocked by plant growth.

- **Evaluating the system's efficiency** periodically by checking soil moisture levels to adapt irrigation schedules appropriately.

Choosing the right irrigation system and managing it effectively are paramount in maximizing water efficiency and ensuring plant vitality. Whether it's for a commercial farm, a residential garden, or extensive landscaping, thoughtful implementation and diligent management of irrigation practices can lead to substantial savings in water and energy. By deeply understanding the environmental conditions, soil characteristics, and specific plant needs, users can craft a strategic approach that supports not only the health of their plants but also contributes to a more sustainable and resilient environment. Through proper planning, design, maintenance, and adaptive management, optimal irrigation practices can be achieved, fostering greater productivity and environmental stewardship.

Book 3: Garden, Crop, and Livestock

Planning and Managing a Self-Sufficient Garden

Planning and managing a self-sufficient garden is a rewarding yet complex endeavor that requires strategic thinking and a hands-on approach. Such gardens are not only a source of fresh produce but also a step towards sustainability and resilience in food supply. This section delves into key aspects of planning, planting, and maintaining a garden that can provide a reliable source of food throughout the year.

Site Analysis

1. **Sunlight Exposure:**

   o Evaluate the daily and seasonal patterns of sunlight across your garden area. Most vegetable crops require a minimum of six hours of direct sunlight to thrive. Observe potential shade sources like buildings, trees, and landscape features that could impact sun exposure.

   o Use tools like a sun calculator or observe the shadow patterns during different times of the day to determine the sunniest spots.

2. **Wind Patterns:**

   o Identify prevailing wind directions and consider how they could affect your plants. Wind can dry out soil quickly, impact pollination processes, and in some cases, can be strong enough to damage plants or erode topsoil.

- o Implement windbreaks such as fences, shrub hedges, or rows of trees strategically placed to reduce wind velocity in the garden area.

3. **Water Sources and Drainage:**

   - o Assess proximity to natural or artificial water sources. Ease of water access is crucial for irrigation purposes, especially in regions with irregular rainfall.

   - o Analyze the soil's drainage capacity by observing how water accumulates or drains after a heavy rain. Poor drainage can lead to waterlogged soil, adversely affecting plant health.

4. **Soil Composition and Topography:**

   - o Slopes can influence water drainage and soil erosion. Gentle slopes are often ideal, but steep inclines may require terracing or other modifications to manage erosion and water flow.

   - o Perform a basic soil test to determine texture, composition, and fertility. This will help in amending the soil appropriately to suit the needs of specific plant types.

Climate and Seasonality

1. **Understanding Local Climate:**

   - o Research the climate zone of your area to determine temperature extremes, average rainfall, and humidity. Climate zones can guide you in selecting plant varieties that are most likely to succeed in your local conditions.

   - o Be aware of microclimates within your property, which may allow for growing plants that wouldn't thrive just a few feet away.

2. **Seasonal Planning:**

   - o Familiarize yourself with the local growing season. The length of the growing season is dictated by the last and first frost dates and will determine when to plant and harvest each crop.

   - o Consider using season extenders like cold frames, greenhouses, or row covers to protect crops from early or late frosts and to potentially extend the growing season.

3. **Plant Selection Based on Climate:**

   - o Choose plant varieties that can adapt to the temperature range and humidity of your area. Some plants may require a cooler climate, while others thrive in heat.

   - o Consider the water needs of each plant in relation to the typical rainfall in your area. Drought-tolerant plants may be necessary in arid regions, whereas water-loving plants will do well in areas with abundant moisture.

Soil Preparation

To establish a thriving and self-sufficient garden, the initial step is to adequately prepare the soil, which is the foundation of garden health and productivity. Here's a detailed exploration of essential soil preparation techniques:

1. **Soil Testing:**

- Conduct a comprehensive soil test before you begin planting. This test will reveal vital information about the nutrient content, pH level, and mineral deficiencies of your soil.
- Testing kits can be purchased from garden centers or online, or you can send a soil sample to a local extension service or laboratory for a more detailed analysis.

2. **Soil Amendments:**

   - Based on the results of your soil test, you may need to adjust the pH or nutrient levels. For acidic soils, lime can be added to increase the pH, while sulfur may be used to decrease the pH of alkaline soils.
   - Incorporate organic amendments like compost, aged manure, bone meal, or green manures. These materials improve soil structure, increase nutrient content, and enhance the soil's water retention and drainage capabilities.

Crop Selection and Rotation

Choosing the right crops and implementing a strategic crop rotation plan are critical for maintaining a healthy garden ecosystem:

1. **Choosing the Right Crops:**

   - Select plants that are suitable for your local climate and soil conditions. Research native species and heirloom varieties, as these are often more resilient and require less supplemental watering and pest control.
   - Consider the growth habits and requirements of each plant. Some plants may require more space or different nutrients than others, which can influence your garden layout and soil preparation.

2. **Crop Rotation:**

   - Crop rotation is a practice used to manage soil fertility and help prevent soil-borne diseases and pest infestations. By rotating crops according to their family and nutrient needs, you can naturally manage the soil's health and reduce the need for chemical interventions.
   - Plan your rotation by dividing crops into groups based on their family (e.g., legumes, brassicas, solanaceae) and nutrient uptake. Rotate these groups in a three or four-year cycle to prevent depleting the soil of specific nutrients and to disrupt the life cycles of common pests.
   - **Example of a Basic Crop Rotation Plan:**
     - **Year 1:** Legumes (beans, peas) – These plants fix nitrogen in the soil, which benefits leafy plants in the following year.
     - **Year 2:** Leafy vegetables (lettuce, spinach) – These are heavy nitrogen consumers.
     - **Year 3:** Root crops (carrots, beets) – These plants are less demanding in terms of nitrogen but require phosphorus.
     - **Year 4:** Fruit-bearing plants (tomatoes, peppers) – These need significant amounts of potassium.

Integrated Pest Management (IPM)

Integrated Pest Management (IPM) is a sustainable approach to managing pests that combines various environmentally friendly strategies to prevent and mitigate pest damage while minimizing the impact on the environment, humans, and non-target organisms. Here's how you can effectively implement IPM in your garden:

1. **Prevention:**

   o   The first line of defense in an IPM strategy involves preventing pests from becoming a threat. This can be achieved by selecting disease-resistant plant varieties and ensuring optimal plant health through proper nutrition and careful spacing to reduce humidity levels around plants.

   o   Maintaining healthy soil is crucial. Rich, well-aerated soil supports strong plant growth, which is inherently more resistant to pests and diseases.

2. **Observation:**

   o   Regular monitoring and early detection are key components of IPM. Inspect your garden frequently for any signs of pest activity or disease. This includes looking for physical damage, such as holes in leaves, discolored spots, or the presence of the pests themselves.

   o   Keeping a garden diary can help track pest appearances and effectiveness of different management strategies over time.

3. **Intervention:**

   o   When pests are detected, IPM does not necessarily resort to chemical pesticides. Instead, options are evaluated in a hierarchy starting with the least possible hazard and risk of impact.

   o   Biological control is preferred, involving the introduction of natural predators or parasites of the pests. For example, ladybugs can be introduced to control aphid populations.

   o   If necessary, targeted applications of organic pesticides may be used as a last resort. These should be selected and applied carefully to minimize their impact on non-target species and the environment.

Water Management

Efficient water management is essential in creating a self-sufficient garden. Here are several techniques that can help conserve water and ensure it is used where most needed:

1. **Drip Irrigation:**

   o   Drip irrigation systems deliver water directly to the base of plants, significantly reducing evaporation and water waste. This system is particularly effective in arid climates or during dry spells.

   o   It can be automated and adjusted according to the specific needs of each plant, which helps in using water efficiently.

2. **Mulching:**

   o   Applying a layer of mulch around plants helps retain soil moisture, suppress weeds, and improve soil quality. Organic mulches such as straw, bark, or leaf clippings also add nutrients to the soil as they decompose.

o   Mulch reduces the frequency of watering needed by minimizing surface evaporation and providing a cooler root environment.

3. **Rainwater Harvesting:**

   o   Collecting rainwater in barrels or tanks from rooftops can provide a significant source of water for gardening. This not only reduces dependency on municipal water systems but also makes good use of a free resource.

   o   Incorporate systems to filter and store rainwater adequately, ensuring it remains clean and safe for garden use.

Harvesting and Storage

Effectively managing the harvesting and storage of garden produce is crucial for maximizing both yield and nutritional value, while also ensuring that you can enjoy the fruits of your labor throughout the year. Here's a comprehensive approach to efficient harvesting and storage practices:

Harvest Timing

- **Peak Ripeness:** Understanding the right time to harvest each type of produce is key. Fruits and vegetables should be picked when they are at their peak of ripeness. This not only ensures optimal flavor and nutritional content but also impacts the longevity of the produce once stored.

- **Regular Harvesting:** Frequent harvesting encourages many plants to increase their yield. For instance, removing ripe vegetables prompts the plant to focus energy on producing new growth.

Storage Solutions

- **Root Cellars:** Utilize root cellars for storing root vegetables like carrots, potatoes, and beets. These underground storage areas naturally provide the cool, humid conditions ideal for prolonging the freshness of these vegetables.

- **Canning:** Preserve a variety of fruits and vegetables through canning. This method involves placing foods in jars and heating them to a temperature that destroys micro-organisms that cause food to spoil.

- **Drying:** Dry herbs, fruits, and vegetables to extend their shelf life. Drying reduces the moisture content in foods, effectively preventing the growth of bacteria, yeast, and mold.

- **Freezing:** Freezing is an excellent way to preserve the taste, nutritional value, and texture of many foods. Almost all fruits, vegetables, and even some herbs can be frozen shortly after harvesting to maintain their quality.

Sustainability Practices

Incorporating sustainability practices into your gardening efforts not only benefits your immediate environment but also contributes to global ecological health. Here are some sustainable techniques to enhance your garden's productivity:

Companion Planting

- **Natural Pest Control:** Certain plant combinations naturally help reduce pest issues. For example, marigolds emit a smell that repels many types of insects, while basil can help ward off thrips and flies when planted near tomatoes.

- **Enhanced Growth:** Companion plants can improve plant health and soil quality. For instance, planting beans near corn can increase nitrogen in the soil, benefiting the corn.

Biodiversity

- **Ecosystem Health:** Increasing the diversity of plant species in your garden supports a wider range of wildlife and promotes a more resilient ecosystem. Diverse plantings can reduce the spread of plant diseases and minimize pest outbreaks.

- **Crop Resilience:** A diverse garden mimics natural ecosystems, reducing the need for chemical inputs and increasing the overall resilience of the garden to climatic stress.

Organic Practices

- **Fertilizers:** Use organic fertilizers such as compost, manure, or bone meal, which release nutrients slowly into the soil, improving its structure and fertility over time without the harsh effects associated with synthetic fertilizers.

- **Pest Control:** Opt for natural pest control methods like introducing beneficial insects, using neem oil, or making DIY organic sprays from household ingredients like garlic and chili pepper. These methods help maintain the natural balance in your garden while keeping it free from harmful chemicals.

Regular Evaluation and Adjustment

Continuously monitor the performance of your garden and be prepared to make adjustments based on what works and what doesn't. Keeping a garden journal can help track plant progress, pest issues, and successful yields, informing future garden plans.

A self-sufficient garden is a dynamic system that requires attention and adaptation. By understanding and responding to the specific needs of your environment, soil, and the crops you grow, you can maximize the productivity and sustainability of your garden. This not only provides a reliable source of fresh, healthy food but also deepens your connection with nature and your understanding of ecological principles.

Advanced Permaculture Techniques

Permaculture is a holistic approach to agriculture that focuses on the conscious design and maintenance of agriculturally productive ecosystems which have the diversity, stability, and resilience of natural ecosystems. It involves arranging plants, animals, and landscape features in ways that mimic the natural ecosystems. Advanced permaculture techniques take these principles further, integrating innovative approaches to create self-sustaining systems. Here's an exploration of some advanced permaculture strategies:

Zoning and Sector Planning in Permaculture

Permaculture is a holistic design system that seeks to mimic the efficiency and sustainability of natural ecosystems. Two of the core components of this design system are zoning and sector planning. These strategies help in optimizing the use of land by carefully considering the interactions between human activity and natural processes. Here is a detailed exploration of each:

Zoning

Zoning is a method of spatial organization in permaculture design that divides the land into zones based on the frequency of human use and the type of activities to be conducted. The zones are numbered from 0 to 5, with each zone having specific characteristics and purposes:

- **Zone 0:** This is the home or central hub where daily activities occur. It is the focal point of human activity, often housing the kitchen garden, tools shed, and compost bins for easy access.

- **Zone 1:** Located just outside the home, this zone is used for plants that require regular attention. Kitchen herbs, salads, and other small plants that are harvested frequently are ideal for this area. It's also a place for more delicate or high-maintenance crops that benefit from close observation and quick access.

- **Zone 2:** This zone houses perennial plants that require less frequent maintenance, such as berry bushes and fruit trees. It may also include larger annuals that need only occasional tending, like pumpkins and squash.

- **Zone 3:** Often used for self-sufficient crops that require minimal maintenance, such as grains and larger fruit trees. This area might also support pastures and livestock that do not need daily attention.

- **Zone 4:** This is a semi-wild area, typically used for foraging and the collection of wild foods. It may also serve as a buffer zone, providing habitat for wildlife and a space for natural ecosystems to function, which can benefit the whole permaculture site.

- **Zone 5:** The wilderness area, this zone is left untouched by active human management and is used for observation of natural ecosystems and processes. It helps in understanding the natural balance and supports biodiversity, which is crucial for the ecological health of the permaculture site.

Sector Planning

Sector planning complements zoning by taking into account the external energies that affect the site. This method identifies and utilizes natural energies and resources such as sunlight, wind, and water:

- **Sun Path:** Understanding the movement of the sun over the area helps in positioning plants and structures to maximize light exposure during critical times of the day and year. This is crucial for passive solar heating and for ensuring that light-dependent plants receive the appropriate amount of sunlight.

- **Wind Patterns:** Identifying prevailing wind directions can assist in placing windbreaks to protect sensitive crops and zones from harsh winds, or in harnessing wind energy.

- **Water Flow:** Analyzing how water moves through the site enables the design of effective water catchment and irrigation systems. Swales, ponds, and other water-harvesting features can be strategically placed to capture and utilize runoff, reducing water waste and maximizing moisture availability.

- **Other Factors:** Other elements like fire risk, noise, and pollution sources are also considered in sector planning, allowing for a design that minimizes negative impacts while enhancing the site's natural resilience and productivity.

Edge Effect

The Edge Effect in permaculture refers to the idea that the greatest productivity and diversity of species occurs at the boundary between two different ecosystems. This concept is derived from ecological observations where edges, such as those between forests and fields, are areas of intense activity and interaction among various species.

- **Maximizing Edges:** To take advantage of the Edge Effect, permaculture designs often incorporate a greater number of edges and transition zones. This can be achieved through the creation of curved beds, using ponds, or planting hedges alongside fields. Each added edge creates a new habitat, new opportunities for niches, and increased edge diversity.

- **Design Techniques:** Utilizing multiple layers in a garden—such as a canopy layer, a shrub layer, and a ground cover layer—can also increase edge effects. This layered approach not only maximizes space but also mimics natural forest systems, creating a multitude of microenvironments where different species can thrive.

- **Benefits:** These techniques enhance the interactions between biological components and the microclimates in those areas, leading to richer, more diverse ecosystems. Edges provide varied light, moisture, and soil conditions, enabling a wider variety of plants and animals to coexist and benefit from each other.

Succession Planting

Succession planting in permaculture is inspired by the natural succession of plants, where ecosystems evolve from a simple, fast-growing pioneer stage to a more complex and stable mature state.

- **Planned Succession:** By understanding and implementing ecological succession, permaculturists can guide the development of their gardens. Starting with quick-growing, nutrient-accumulating species that prepare the soil, they gradually introduce more permanent plants that require richer conditions to thrive.

- **Implementation:** For example, a newly established garden might begin with legumes that fix nitrogen in the soil, followed by fast-growing vegetables, and eventually transition to perennial plants and trees. This method mimics natural processes, which can significantly reduce maintenance needs and enhance the garden's resilience.

- **Strategic Planning:** Planned succession helps in creating a self-regulating garden. As the garden matures, the ecological balance established reduces the need for external inputs such as fertilizers and pest control, as the system begins to maintain itself.

- **Benefits:** This approach not only ensures continuous production throughout the growing season but also builds soil health, increases biodiversity, and leads to a more sustainable and self-sufficient garden environment.

Aquaponics and Aquaculture: Integrated Water Systems for Sustainable Agriculture

Aquaponics and aquaculture are innovative agricultural practices that merge the principles of hydroponics (soilless plant cultivation) with those of aquaculture (fish farming). These systems are designed to maximize efficiency, sustainability, and productivity in both urban and rural settings.

Aquaponics: A Synergistic System

Aquaponics is an integrated approach where the waste produced by farmed fish or other aquatic animals supplies nutrients for plants grown hydroponically, which in turn purify the water that returns to the fish environment.

- **Closed-loop Sustainability:** In an aquaponics system, fish produce ammonia-rich waste. This waste is converted by bacteria into nitrates, which are excellent nutrients for plant growth. The plants absorb these nutrients, which cleans the water that goes back into the fish tanks. This cyclical system mimics natural ecological cycles, reducing the need for chemical inputs and creating an efficient loop of crop and protein production.

- **Space and Resource Efficiency:** Aquaponics systems can be set up in a variety of spaces, ranging from small indoor units to larger greenhouse setups. They use up to 90% less water than traditional agriculture by recycling water continuously. Additionally, these systems do not require arable land and can produce high yields of both plants and fish in compact areas, making them ideal for urban environments where space and water might be scarce.

- **Diversity of Produce:** These systems are not only efficient but also versatile, capable of growing a wide range of crops—from leafy greens like lettuce and herbs to fruit-bearing plants like tomatoes and peppers—alongside a variety of fish, such as tilapia, trout, and ornamental species.

Aquaculture: Enhancing Biodiversity and Productivity

Aquaculture involves cultivating aquatic organisms in controlled environments, focusing on sustainable practices to raise fish and other aquatic species.

- **Biodiversity and Ecosystem Benefits:** By integrating aquaculture ponds or tanks into agricultural systems, farmers can enhance on-site biodiversity and create more resilient ecosystems. Aquatic systems can act as habitats for various species, thereby supporting overall ecological health.

- **Multiple Yields:** Aquaculture can produce not only fish but also other products such as algae and aquatic plants. These can serve multiple purposes—providing food for human consumption, feed for animals, and even organic fertilizer for crops. Additionally, water from aquaculture systems can be used to irrigate fields, bringing nutrients directly to terrestrial crops.

- **Sustainability Practices:** Modern aquaculture systems are increasingly adopting eco-friendly practices such as recirculating systems, which minimize water use and pollution. There is also a growing emphasis on raising native and herbivorous species that require lower levels of input and are less likely to cause ecological disruption if they escape into the wild.

Aquaponics and aquaculture represent progressive steps towards solving some of the pressing issues in agriculture such as water scarcity, land degradation, and the need for sustainable protein sources. By combining these systems, practitioners can not only boost their productivity but also contribute to a more sustainable and ecologically sound agricultural paradigm.

Polyculture and Crop Diversity: Enhancing Agricultural Resilience

Polyculture and crop diversity are key elements of sustainable agriculture, focusing on the cultivation of multiple crop species within the same area. This approach not only mimics natural ecosystems but also significantly enhances biodiversity, resilience, and productivity of agricultural systems.

Polyculture Systems: Multifunctional Agriculture

Polyculture involves growing a variety of crops in close proximity, in contrast to the monoculture systems that dominate much of modern agriculture. This method offers several profound benefits:

- **Increased Biodiversity:** By cultivating a range of species, polyculture systems support a wider variety of beneficial insects and soil organisms, which contribute to a healthier ecosystem. Increased plant diversity leads to more complex and stable ecosystems that can better resist pests and diseases.

- **Natural Pest Management:** In a diverse planting scheme, pests are less likely to find and devastate a particular crop, as the presence of multiple plant species creates a more confusing environment for pests.

Additionally, the increased biodiversity attracts natural predators that help keep pest populations in check.

- **Improved Soil Health:** Different plants have varying root structures and nutrient needs, which can help improve soil structure and fertility over time. For example, deep-rooted plants can bring nutrients up from lower soil layers, while nitrogen-fixers can naturally enrich the soil.

- **Enhanced Yield Per Area:** Polyculture systems can produce more food per unit area as plants with different growth habits and requirements can effectively share space. For instance, tall plants provide shade for those requiring less sun, while ground cover crops suppress weeds and preserve soil moisture.

Guild Planting: Synergistic Plant Communities

Guild planting takes the concept of polyculture a step further by intentionally grouping plants that interact beneficially with each other. This mimics the cooperative relationships found in natural ecosystems.

- **Supportive Relationships:** Plants in a guild support each other in various ways—some might provide necessary shade, others might repel harmful insects, and some can fix nitrogen in the soil to the benefit of their neighbors. For example, a classic guild includes corn, beans, and squash—often referred to as the "Three Sisters"—where each plant contributes to the success of the others.

- **Attracting Beneficial Insects:** Some plants can attract specific beneficial insects that not only pollinate the crops but also control pest populations. For example, flowers like marigolds or sunflowers planted near vegetable crops can attract bees and other pollinators, as well as beneficial predators that feed on common pests.

- **Resource Optimization:** Guild planting allows for the efficient use of water, light, and nutrients. Different species have varying requirements, and by carefully planning their placement, farmers can ensure that each plant gets what it needs without competing too harshly with its neighbors.

Polyculture and guild planting are effective strategies for sustainable agriculture, promoting ecological balance and resource efficiency. By adopting these practices, farmers and gardeners can create more productive, resilient, and environmentally friendly agricultural systems

Soil Building and Rehabilitation: Enhancing Soil Health and Fertility

Soil building and rehabilitation are central to sustainable agriculture, particularly within the framework of permaculture, which seeks to create ecologically harmonious, efficient, and productive systems. A focus on nurturing the soil food web and employing rehabilitation techniques is crucial to these efforts, as healthy soil is the foundation of a thriving ecosystem.

Soil Food Web Enhancement

The soil food web refers to the complex network of organisms living within the soil, including bacteria, fungi, protozoa, nematodes, earthworms, and many other types of life. These organisms play critical roles in the decomposition of organic matter, nutrient cycling, and the overall health of soil ecosystems. Here are some strategies used in advanced permaculture to support the soil food web:

- **Composting:** Composting is a vital process in soil management, converting organic waste into rich humus. This not only recycles nutrients but also introduces beneficial microorganisms into the soil, enhancing its structure and fertility.

- **Mulching:** Applying organic mulch, such as straw, leaves, or wood chips, helps conserve moisture, suppress weeds, and improve soil quality. As the mulch decomposes, it provides ongoing nutrition to soil microorganisms, which in turn benefit plant health.

- **Integrating Livestock:** Animals play a natural role in the soil food web by grazing, manuring, and trampling. This activity helps cycle nutrients, aerate the soil, and integrate organic matter into the soil matrix, fostering a richer, more diverse microbial ecosystem.

Rehabilitation Techniques for Degraded Soils

Rehabilitating degraded soils is a challenge that requires innovative and effective strategies. Permaculture offers several approaches to rejuvenate and enhance soil health:

- **Dynamic Accumulators:** Certain plants, known as dynamic accumulators, have deep root systems that can tap into subsoil nutrients unavailable to other plants. Examples include comfrey, dandelion, and yarrow. These plants draw up minerals like potassium, calcium, and magnesium, which are then deposited on the soil surface as the plants shed leaves or are chopped and dropped as mulch.

- **Cover Cropping:** Growing cover crops such as clover, vetch, and rye helps improve soil structure, prevent erosion, and fix nitrogen—a crucial element for plant growth. Cover crops can be tilled into the soil, adding organic matter and nutrients.

- **Biochar Application:** Incorporating biochar, a form of charcoal produced from plant matter, into the soil can improve its fertility and ability to retain water and nutrients. Biochar also provides a habitat for soil microorganisms, enhancing the soil food web.

- **No-Till Farming:** Minimizing or eliminating plowing and tilling protects soil structure, prevents erosion, and maintains the integrity of the soil food web. No-till methods increase organic matter retention and foster a stable soil environment conducive to plant growth.

By implementing these advanced permaculture techniques, practitioners can create robust, sustainable ecosystems that require fewer inputs, produce more outputs, and sustain a diverse array of flora and fauna. This not only benefits the environment but also provides a blueprint for sustainable living and farming practices that can be replicated in diverse conditions around the world.

Responsible and Sustainable Livestock Farming

Responsible and sustainable livestock farming is essential for creating a resilient food system that is both eco-friendly and economically viable. This approach to livestock farming emphasizes the health of the animals, the environment, and the community, aiming to produce meat, dairy, and eggs in ways that can sustain future generations without depleting resources or harming the planet.

1. Animal Welfare

The ethical treatment of livestock is a cornerstone of responsible farming. Ensuring animal welfare involves several critical practices that contribute to the health and productivity of the animals, which, in turn, affect the quality of the products they produce.

- **Living Conditions**: Providing spacious and comfortable living conditions is essential. Animals should have enough room to move freely and engage in natural behaviors. This reduces stress and aggression among animals and promotes healthier living conditions.

- **Appropriate Feed**: Animals must receive a diet that meets their nutritional needs at different stages of their life. This includes a balance of grains, proteins, and fibers, tailored to their specific dietary requirements. The use of high-quality, non-GMO, or organic feed can improve their health and the quality of the meat, milk, or eggs they produce.

- **Regular Medical Care**: Preventive health care, including vaccinations, routine check-ups, and prompt treatment of illnesses, ensures that animals remain healthy and productive. This also includes regular monitoring for signs of disease or distress.

- **Enrichment**: Providing environmental enrichment, such as scratching areas for pigs, dust baths for chickens, or sufficient grazing land for cattle, can significantly improve an animal's quality of life and overall wellbeing.

2. Pasture Management

Effective pasture management is vital in sustainable livestock farming, helping to prevent overgrazing, maintain healthy soil, conserve water, and support biodiversity.

- **Rotational Grazing**: Implementing rotational grazing systems where animals are moved between different pasture areas allows grasses and other vegetation time to recover, regrow, and maintain their health. This technique mimics the natural movement patterns of wild grazing animals, which is beneficial for both the land and the livestock.

- **Soil Health**: Rotational grazing also helps in maintaining soil structure and fertility. By preventing overgrazing, the soil is less likely to become compacted and eroded, preserving its ability to absorb and retain water and nutrients.

- **Water Management**: Efficient management of water resources on pastures is crucial. Ensuring that water sources are clean and accessible improves animal health and helps maintain the ecological balance of the grazing areas.

- **Biodiversity Enhancement**: Diverse pastures that include a variety of grasses, legumes, and other plants not only provide better nutrition to the grazing animals but also support an array of wildlife species. Increasing plant diversity can attract beneficial insects and promote a healthier ecosystem.

3. Feed Efficiency

Improving feed efficiency is a critical aspect of sustainable livestock management. It involves optimizing the feed used for livestock to ensure maximum nutritional benefits while minimizing waste. This practice not only reduces the cost of feed but also lessens the environmental impact associated with livestock farming.

- **Selection of Appropriate Diets**: Choosing the right feed involves understanding the specific dietary needs of different types of livestock based on their age, weight, health, and production goals (such as milk, meat, or eggs). High-quality diets improve digestion and nutrient absorption, reducing the amount of feed needed and the waste produced.

- **Use of Locally Sourced, Organic Feeds**: Incorporating organic feeds from local sources decreases the reliance on imported feeds, which often have a higher carbon footprint due to transportation. Local sourcing supports local agriculture and economies, and organic feeds ensure that livestock are not consuming artificial additives or genetically modified organisms, appealing to health-conscious consumers.

- **Feed Supplements**: Adding supplements to enhance the nutritional value of the feed can improve animal health and productivity. Supplements like enzymes, amino acids, and probiotics can significantly boost feed efficiency, improving growth rates and production.

4. Manure Management

Manure management is a key component of environmentally responsible livestock farming. Proper handling and processing of manure can mitigate environmental impacts and turn a potential waste product into a valuable resource.

- **Composting**: Composting manure converts it into a stable, nutrient-rich organic fertilizer that can be used to enrich the soil in crop production. This process reduces odors, destroys pathogens, and decreases the volume of waste, making it easier to handle and use effectively.

- **Anaerobic Digestion**: This process involves breaking down organic matter in the absence of oxygen to produce biogas, which can be used as a renewable energy source. The residue, known as digestate, is an excellent fertilizer, rich in nutrients. Anaerobic digestion reduces greenhouse gas emissions, particularly methane, which is a potent greenhouse gas often associated with livestock farming.

- **Integrated Nutrient Management**: By integrating manure as a fertilizer within the farm's overall nutrient management strategy, farmers can reduce their dependence on chemical fertilizers, which are energy-intensive to produce and contribute to nutrient runoff that can pollute waterways.

- **Regulatory Compliance**: Effective manure management also helps livestock farmers comply with environmental regulations, which can restrict the amount of nutrient runoff and mandate the proper treatment and disposal of livestock waste to protect water quality and soil health.

5. Integrated Farming Systems

Integrated farming systems represent a holistic approach to agriculture that combines livestock and crop production in a mutually beneficial cycle. This strategy enhances the efficiency of resource usage and promotes a sustainable agricultural model.

- **Nutrient Cycling**: Livestock contribute essential nutrients to the soil through their manure, which can be used to fertilize crops naturally. This recycling of nutrients reduces the need for chemical fertilizers, which are often costly and environmentally damaging.

- **Crop-Livestock Synergy**: Animals can feed on crop residues that might otherwise be wasted, such as stalks and husks, while the crops can provide high-quality fodder to sustain the animals. This reciprocal relationship maximizes the use of available resources, reducing waste and increasing farm productivity.

- **Diversification of Farm Income**: By diversifying farm operations to include both crops and livestock, farmers can reduce risk and increase economic stability. Crop failures may be offset by livestock sales, and vice versa, providing a financial buffer that can sustain the farm's operations through variable market and weather conditions.

- **Enhancing Ecological Resilience**: The integration of crop and livestock farming can increase biodiversity on the farm, which in turn enhances resilience to pests and diseases. A diverse agricultural ecosystem is more likely to withstand environmental stressors and less dependent on external inputs.

6. Water Conservation

Water management is a critical aspect of sustainable livestock farming, particularly in areas susceptible to drought or where water resources are limited.

- **Rainwater Harvesting**: Implementing rainwater harvesting systems can capture rainfall that can be used for livestock watering, cleaning, and irrigation. This not only conserves precious groundwater and surface water resources but also reduces the farm's operational costs.

- **Recycling Gray Water**: Gray water, which includes water from baths, sinks, and other domestic sources, can be treated and reused for agricultural purposes. Using gray water for livestock can significantly reduce the demand for fresh water and help maintain the water cycle within the farm.

- **Protecting Water Sources**: Ensuring that livestock farming does not contaminate local water bodies is crucial. Implementing buffer zones, proper waste disposal practices, and well-designed water distribution systems can prevent runoff of nutrients and pathogens into watercourses, protecting ecosystem health and water quality.

- **Efficient Water Use Practices**: Techniques like precision watering, where water use is strictly managed and tailored to the needs of both crops and livestock, can minimize waste. Techniques such as drip irrigation for crops and water trough innovations for animals can significantly improve water efficiency.

7. Genetic Diversity in Livestock

Genetic diversity within livestock populations is crucial for resilience, productivity, and adaptability to changing environmental conditions and emerging diseases. Preserving a broad genetic base can provide the following benefits:

- **Disease Resistance**: Diverse genetics often enhance an animal's natural defenses against diseases, reducing the need for medical interventions such as antibiotics, which can be costly and have ecological implications.

- **Adaptability**: Animals with diverse genetic backgrounds are more likely to adapt to varying climatic conditions and shifts in habitat. This adaptability is crucial in the face of climate change, where temperatures and weather patterns are increasingly unpredictable.

- **Preservation of Heritage Breeds**: Many heritage and rare breeds possess unique traits that are invaluable for future breeding strategies and maintaining biodiversity. These breeds often have qualities such as drought resistance, disease resistance, and the ability to thrive on marginal pastures.

- **Sustainable Breeding Practices**: Implementing breeding programs that focus on enhancing genetic diversity can lead to healthier and more productive livestock. Sustainable breeding includes selecting traits that improve longevity, fertility, and overall animal welfare.

8. Reducing Carbon Footprint

Reducing the carbon footprint in livestock farming is essential for mitigating climate change and promoting environmental sustainability. Effective strategies include:

- **Optimizing Transportation**: Streamlining transport logistics to minimize the distance food travels from farm to consumer reduces greenhouse gas emissions. This can be achieved by localizing feed sources and markets, thus reducing dependency on long-haul transport.

- **Enhancing Feed Conversion Ratios**: Improving the efficiency of feed conversion in livestock reduces the amount of feed required, thereby decreasing the production of methane from enteric fermentation, a significant source of emissions in ruminant animals.

- **Employing Renewable Energy Sources**: Incorporating renewable energy technologies such as solar panels, wind turbines, or biogas systems can significantly reduce a farm's reliance on fossil fuels. These technologies not only cut emissions but also can lead to substantial savings in energy costs over time.

- **Carbon Sequestration Practices**: Techniques such as managed grazing can enhance the ability of pastures to sequester carbon. By maintaining healthy grasslands and optimizing grazing patterns, farmers can help capture atmospheric carbon dioxide and store it in the soil.

Implementation Strategies for Sustainable Livestock Farming

Education and Training

Proper education and training are fundamental in equipping farmers with the necessary skills and knowledge to implement sustainable livestock practices. Key approaches include:

- **Agricultural Extension Services**: These services play a crucial role in bridging the gap between research and practical application. Extension agents can provide one-on-one advice, deliver the latest research findings, and demonstrate sustainable farming techniques directly on farms.

- **Workshops and Seminars**: Organizing regular educational workshops and seminars can help keep farmers updated on new methods, technologies, and regulatory changes affecting sustainable livestock management.

- **Field Demonstrations**: Practical demonstrations on farms that already implement sustainable practices can provide a real-world example of how these methods work and their benefits, encouraging adoption among other farmers.

Community Engagement

Community engagement is vital in creating a supportive environment for sustainable livestock farming. Strategies include:

- **Promotion of Locally Produced Goods**: By engaging the local community, farmers can increase awareness about the benefits of buying local and sustainably produced livestock products. This not only supports the local economy but also reduces the environmental impact associated with long-distance food transportation.

- **Community-Supported Agriculture (CSA) Programs**: CSA programs allow consumers to buy shares of a farm's production, giving them a regular supply of fresh products while providing farmers with a stable income and a direct market for their goods.

- **Educational Outreach**: Hosting open farm days, school visits, and community workshops can increase public awareness about the importance of sustainable practices and the challenges faced by modern farmers.

Policy and Incentives

Government policies and incentives are critical to fostering sustainable practices in livestock farming by making them economically viable and attractive to farmers. Effective measures might include:

- **Subsidies and Grants**: Financial incentives such as subsidies for organic farming or grants for purchasing sustainable farming equipment can reduce the initial cost burden for farmers transitioning to sustainable practices.

- **Regulatory Frameworks**: Implementing strict regulations that enforce sustainable practices can ensure that all livestock operations adhere to environmental standards. Penalties for non-compliance can act as a deterrent against unsustainable practices.

- **Support for Renewable Energy**: Providing incentives for renewable energy solutions in agriculture, such as solar or biogas, can encourage farmers to reduce their carbon footprint and energy costs.

- **Research and Development Funding**: Investing in research aimed at improving sustainable livestock management techniques can lead to innovations that make these practices more effective and appealing to farmers.

Responsible and sustainable livestock farming not only addresses ethical concerns regarding animal welfare and environmental conservation but also provides a viable economic model for farmers. By adopting these practices, farmers can ensure that they are part of a sustainable food system that respects the earth and its inhabitants.

Book 4: Off-Grid DIY Tools and Technology

Techniques for Crafting Essential Tools in Off-Grid Living

Crafting essential tools while living off-grid is both a practical skill and a necessity for self-sufficiency. This chapter explores the techniques and methods required to create and maintain tools using materials and resources available in an off-grid environment.

Material Sourcing for Off-Grid Tool Crafting

Sourcing materials for tool crafting is a fundamental skill for off-grid living. This segment explores how to effectively gather and utilize both natural and reclaimed materials to make tools that are both functional and sustainable.

Natural Materials

Utilizing natural materials is not only environmentally friendly but also efficient in an off-grid setting where commercial resources might be limited.

- **Wood**: One of the most versatile and accessible materials in natural settings. Hardwoods such as oak and hickas are prized for their strength and durability, making them ideal for tool handles and structural components. Softer woods may be used for less demanding applications like crafting stakes or utensils.

- **Stone**: Flint and obsidian are excellent for creating sharp edges and have been used in tool making for millennia. These materials can be knapped (shaped by striking) into fine blades for cutting or carving. Other types of stone, like granite or sandstone, can be shaped into grinding tools or building blocks.

- **Bone and Antler**: These materials are useful for more than just aesthetic purposes; they can be carved into needles, awls, and even gears or toggles. Bones and antlers have natural strength and flexibility, which are advantageous in crafting specialized tools like fishing hooks or detailed carvings.

Reclaimed Materials

Repurposing discarded materials is a cornerstone of sustainable living, particularly relevant in off-grid environments where every resource is valuable.

- **Metals from Scrap**: Old machinery, vehicles, and appliances can be treasure troves of metal that can be repurposed. Steel can be reforged into new tools, while copper and aluminum can be melted down and reshaped into fittings, wires, and even small engine components.

- **Plastic and Rubber**: While not traditionally associated with 'natural' off-grid living, salvaged plastics and rubber can be incredibly useful. Plastic containers can be repurposed into parts for water systems or as molds for casting metals. Rubber from tires can be cut and used for waterproofing or as gaskets.

- **Fabric and Fibers**: Old clothing and fabric can be reworked into useful items like bags, protective covers, or even insulation. Fibers can be extracted and twisted into rope or used as binding material.

Techniques for Processing and Using Sourced Materials

- **Processing Wood**: Seasoning wood (allowing it to dry) before use in tool making is crucial to prevent warping. Techniques such as carving, splitting, and sanding are necessary to prepare wood for specific uses.

- **Metalworking**: Basic forging to reshape metal involves heating and hammering it into desired forms. More advanced techniques might include welding or machining, depending on the available equipment.

- **Stone Shaping**: Using harder stones to chip away at softer ones allows for the creation of specific shapes and edges, which is a basic form of knapping.

Sustainable Practices

- **Selective Harvesting**: When gathering natural materials, it's important to do so in a way that doesn't harm the environment. For example, only taking fallen wood or harvesting stone from areas where it won't cause erosion.

- **Recycling and Reusing**: Always look for opportunities to reuse materials. This not only conserves resources but also reduces the waste produced in an off-grid living situation.

Tool Making Techniques for Off-Grid Living

Crafting tools from scratch is a foundational skill for anyone living off-grid. This section details essential techniques in woodworking, metalworking, and stone carving that enable you to create tools and items necessary for daily life without relying on manufactured goods.

Woodworking

Woodworking is more than just a craft; it's an essential skill for building everything from basic furniture to complex machinery components. Here's how you can master woodworking off-grid:

- **Carving and Whittling**: These are the most basic forms of woodworking, useful for creating detailed shapes and designs in wood. Using a knife or a chisel, wood can be shaped into handles, spoons, or artistic sculptures.

- **Joinery**: This involves creating joints that hold pieces of wood together without the need for nails or screws. Techniques such as dovetails, mortise and tenon, or lap joints are not only aesthetically pleasing but also provide strong, long-lasting connections.

- **Finishing**: Proper finishing techniques, like sanding and sealing, are crucial to protect wooden tools and furniture from the elements, particularly in outdoor settings.

Metalworking

Metalworking is indispensable for creating more durable and complex tools.

- **Basic Blacksmithing**: Setting up a forge, even a rudimentary one, allows you to heat and shape metal. This process is vital for making knives, hoes, shovels, and other metal tools that require durability and precision.

- **Bending and Hammering**: These techniques are used to alter the shape of the metal into usable tools and components. For example, bending rods into brackets or hammering sheets into blades.

- **Welding**: While more advanced, if you have access to a welder or can create a makeshift one, welding can significantly expand the types of projects you can undertake, such as repairing tools or constructing metal frameworks.

Stone Carving

Stone tools are not just relics of the past but can be integral to off-grid living, providing you with resources for cutting and building.

- **Flaking**: This technique involves striking a piece of stone with another harder item (usually a hammerstone) to remove small flakes and shape the stone into a sharp edge. This is useful for making simple cutting tools or arrowheads.

- **Grinding and Polishing**: After flaking, stone edges can be refined and sharpened by grinding against a softer stone or sandstone. Polishing can then be done with leather or fabric to smooth out the surface for a more finished look.

- **Drilling**: With patience and the right techniques, stone can also be drilled to create holes for wooden handles or to help in constructing larger structures.

Practical Applications

- **Tool Repair and Maintenance**: Beyond making new tools, these skills are also crucial for maintaining and repairing existing tools, which is a necessary part of sustainable living.

- **Creative Solutions**: Often, off-grid situations require improvisation. Combining these skills can lead to innovative solutions, such as creating a windmill from wood and metal for pumping water.

Tool Maintenance for Off-Grid Living

Maintaining tools is just as critical as making them, especially in an off-grid environment where access to replacements can be limited. Proper maintenance not only extends the life of tools but also ensures they remain effective and safe to use. Here are essential maintenance practices for keeping your tools in top condition:

Sharpening

Keeping tools sharp is fundamental to their performance. Dull tools are not only inefficient but also dangerous, as they require more force to use, increasing the risk of accidents.

- **Using a Whetstone**: For knives, axes, and other bladed tools, a whetstone is ideal for sharpening. Soak the stone in water (or oil, depending on the type of stone), and use a circular or back-and-forth motion across the blade to hone the edge.

- **Files and Sandpaper**: For larger tools like shovels or hoes, a metal file is suitable for removing nicks and restoring a sharp edge. Sandpaper can be used for finer sharpening and smoothing.

Handling

The handles of tools are critical for safety and usability. A broken or unstable handle can render a tool useless or dangerous.

- **Replacing Handles**: Wooden handles on tools like hammers or axes can crack or break. Replacing them involves selecting the right type of wood and shaping it to fit securely.

- **Tightening Components**: Regularly check nuts, bolts, and other fasteners on tools to ensure they are tight. Loose components can lead to poor performance or safety hazards.

Rust Prevention

Rust can significantly degrade the quality and lifespan of metal tools, making rust prevention a necessary aspect of tool maintenance.

- **Oiling Metal Surfaces**: Applying a light layer of oil to metal parts after each use can prevent rust by creating a barrier against moisture. Common oils used include vegetable oil, motor oil, or specialized tool oils.

- **Beeswax or Lard Coating**: For tools that are not used frequently, applying a coat of beeswax or lard can provide long-lasting protection against oxidation. This method is particularly useful for items like garden shears or saws that may be stored for longer periods.

- **Proper Storage**: Store tools in a dry, protected place to prevent exposure to moisture. Tool sheds or boxes with desiccants to absorb moisture can help keep tools dry and rust-free.

Regular Inspection

- **Wear and Tear**: Regularly inspect tools for signs of wear or damage. Check for cracks in wooden handles, warping, or metal fatigue.

- **Cleaning**: After each use, tools should be cleaned to remove dirt, sap, or other materials that might cause damage over time. This is especially important for gardening tools that come into contact with soil and plant material.

Custom Tool Creation for Off-Grid Living

Off-grid living challenges residents to think creatively about how they use and create tools, often leading to innovations that tailor solutions to specific environmental and situational needs. Here's how you can develop and implement custom tool solutions:

Innovating Solutions

The need for specific tools that address unique problems can inspire inventive solutions that are not available commercially. These custom tools can significantly enhance efficiency and effectiveness in daily tasks.

- **Multi-functional Tools**: Designing a tool that combines several functions into one can save space and increase its utility. For example, a multi-tool that serves as a hatchet, wire cutter, and hammer can be invaluable in an environment where versatility and storage space are at a premium.

- **Specific Adaptations**: Consider the specific challenges of your environment—such as thick underbrush or heavy snow. Crafting tools like a bush-clearing machete or a compact, efficient snow shovel can make daily tasks much easier and less time-consuming.

Modular Tools

Modular tools with interchangeable parts can greatly enhance the versatility of your toolkit without occupying much space. This approach allows for easy customization and adaptation to various tasks with minimal resources.

- **Interchangeable Heads**: Create a system where multiple tool heads attach to a single handle. This could include a shovel head, a rake, a hoe, and perhaps even specialized gardening tools like a cultivator or a dibble for planting. Such a system reduces the need to store multiple bulky handles and makes it easier to carry tools around your property.

- **Quick-Change Mechanisms**: Develop or adapt mechanisms that allow for quick and easy changes between tools. This could involve simple, robust locking devices that can withstand the rigors of heavy use without failing.

Custom Fabrication

The ability to fabricate and modify tools according to your specific needs is a valuable skill in off-grid living.

- **DIY Workshops**: Set up a basic workshop that includes the tools necessary for cutting, shaping, and assembling various materials—such as metal, wood, and plastic. This space can also serve as a creative hub where new ideas and tools are born.

- **Utilizing Local Resources**: Use materials that are readily available in your surroundings. For instance, if you have access to a forest, wood can be your primary material for tool handles, while scrap metal from old vehicles or machinery can be repurposed into tool heads or blades.

Community Collaboration

Engaging with a community of like-minded individuals can lead to shared innovations and solutions.

- **Skill Sharing**: Participate in or organize workshops where community members can share their knowledge and skills in tool-making. This can lead to the development of new ideas and the improvement of existing tools.

- **Tool Swaps**: Organize events where tools and ideas can be exchanged. This not only fosters community spirit but also provides access to a wider variety of tools and solutions.

Mastering the art of tool crafting is a key component of successful off-grid living. It enhances self-reliance and empowers individuals to solve problems creatively and efficiently. By developing skills in material sourcing, tool making, and maintenance, off-gridders can ensure they have the tools necessary to build and sustain their lifestyle.

Building Simple Machines for Agriculture and Woodworking for Off-Grid Use

Off-grid farming and woodworking demand efficient, reliable tools and machines that can be maintained with minimal access to modern facilities. Building simple machines for these purposes not only enhances productivity but also aligns with the sustainable principles of off-grid living. Here's how to approach building basic yet effective machinery for agricultural and woodworking tasks:

Agriculture Machinery for Off-Grid Use

1. Manual Seed Planter

A manual seed planter is an invaluable tool for sowing seeds quickly and uniformly without the physical strain associated with traditional hand planting.

- **Design and Materials**:
  - **Frame**: Construct the frame using lightweight aluminum pipes, which are durable yet easy to handle. Assemble the frame in a configuration that allows it to be easily pushed or pulled across the field.
  - **Wheels**: Attach old bicycle wheels at either end of the frame. These wheels will help the planter move smoothly over the soil and make the tool easier to navigate.
  - **Seed Dispenser**: Create a seed dispenser using a combination of small tubes or funnels that can guide the seeds to the soil at consistent intervals. Attach a rotating mechanism that drops a single seed at a time as the wheel turns.
- **Benefits**:
  - Provides consistent seed depth and spacing, ensuring better crop germination and growth.
  - Reduces back strain and fatigue from bending over to plant seeds manually.

2. Hand-Powered Thresher

A hand-powered thresher can efficiently separate grains from their stalks, enabling faster processing of crops like wheat and rice.

- **Design and Materials**:
  - **Frame**: Build a sturdy frame from wood. This frame should support a rotating drum or cylinder equipped with beaters or brushes.
  - **Threshing Mechanism**: Attach metal beaters or brushes to a cylindrical core that can be turned with a hand crank. As the cylinder spins, it should effectively beat the grains from the stalks.
  - **Collection Area**: Ensure there is a method to collect the separated grains, such as a bin or bag positioned below the threshing cylinder.
- **Benefits**:
  - Allows for quick processing of large quantities of grain without the need for electricity.
  - Significantly reduces the labor and time involved in threshing by hand.

3. Wheel Hoe

The wheel hoe is a versatile tool that aids in weeding and tilling, ideal for managing large garden plots.

- **Design and Materials**:
  - **Wheel**: Use a sturdy, large diameter wheel that can handle rough terrain. This could be repurposed from an old cart or wheelbarrow.
  - **Hoe Blade**: Attach a sharp metal hoe blade just behind the wheel. Ensure the blade is replaceable as it will wear out with use.
  - **Handle**: Fit a long wooden handle to the frame, allowing the user to operate the hoe while standing. This handle should be securely attached and comfortable to grip.
- **Benefits**:

- o Efficiently cuts through soil and weeds, reducing the need for multiple passes.

- o Minimizes strain on the back and arms, as the wheel assists in forward motion and the blade placement requires less force to penetrate the soil.

Woodworking Machines for Off-Grid Use

For individuals living off-grid, crafting their own woodworking machines can significantly enhance their ability to process wood for various uses, from building to crafting. Here are detailed instructions on how to build three basic but essential woodworking machines using simple tools and materials.

1. Foot-Powered Lathe

A foot-powered lathe enables precise wood turning without the need for electrical power, making it ideal for crafting bowls, spindles, and other detailed wooden objects.

- • **Design and Materials**:

  - o **Frame**: Construct a robust frame from hardwood or metal to ensure stability and durability. The frame should be heavy enough to withstand the force of turning without wobbling.

  - o **Drive Mechanism**: Install a foot pedal connected to a flexible drive shaft. The pedal powers the lathe's rotation, allowing for hands-free operation. A system of belts and pulleys can be used to transfer motion from the pedal to the spindle.

  - o **Spindle**: Attach a spindle to the frame, where the wood piece will be mounted. Ensure it is perfectly aligned for accurate turning.

- • **Benefits**:

  - o Allows for precision in woodworking, enabling the creation of symmetrical and smooth wooden pieces.

  - o Reduces the need for electrical power, perfect for off-grid living.

2. Wood Splitter

A manual wood splitter can greatly reduce the time and effort required to prepare firewood or split logs for other uses.

- • **Design and Materials**:

  - o **Frame**: Use a sturdy iron or steel frame to support a weighted blade. The frame should include a long arm to which the blade is attached.

  - o **Blade Mechanism**: The blade should be heavy and sharp, mounted on a pivot at the top of the frame. Users lift the blade by the arm and drop it to split wood placed below.

  - o **Base**: Ensure the base is stable and can securely hold wood logs in place during splitting.

- • **Benefits**:

  - o Significantly reduces the physical effort and time needed to split wood compared to traditional axe splitting.

o Provides a safer alternative to swinging an axe, minimizing the risk of injury.

3. Saw Bench

A saw bench is a crucial tool for any woodworker, providing a stable platform for precise cutting.

- **Design and Materials**:

  o **Construction**: Build the bench from solid wood, ensuring it is sturdy enough to withstand vigorous sawing. The bench should be at a comfortable height to allow easy handling of materials.

  o **Clamps and Grips**: Incorporate clamps or grips into the design to securely hold wood in place while it is being sawed. These should be adjustable to accommodate different sizes and shapes of wood.

  o **Safety Features**: Add safety features such as a non-slip surface and perhaps guards around any exposed blades or cutting mechanisms.

- **Benefits**:

  o Provides a secure and stable surface for cutting wood, improving both safety and accuracy.

  o Enhances efficiency in woodworking projects by allowing for quicker and more precise cuts.

Design Principles for Simple Machines for Off-Grid Use

Creating simple machines for use in off-grid environments involves a balance of durability, simplicity, efficiency, and sustainability. These principles ensure that the tools are practical for their intended uses while being environmentally considerate and manageable to maintain. Here's how to incorporate these principles into the design and construction of simple machines:

Durability

- **Materials**: Choose materials that are tough and able to endure harsh conditions. Treated wood resists rot and pest damage, while metals like stainless steel and aluminum offer resistance to rust and degradation. These materials ensure longevity and reduce the need for frequent replacements.

- **Construction**: Build machines to be robust with reinforced joints and protection against elements. For example, use waterproof seals and rust-proof coatings to enhance durability.

Simplicity

- **Design**: Construct machines with minimal complexity to ease the understanding and operation by any user. Simple designs are easier to repair and less likely to break down because they have fewer parts that can fail.

- **Maintenance**: Ensure that all parts of the machine are easily accessible for cleaning, repairs, and replacement. Using standard, interchangeable parts can help make repairs more straightforward.

Efficiency

- **Leverage Mechanical Advantage**: Incorporate basic mechanical principles such as levers, pulleys, and inclined planes to minimize the effort required to perform tasks. For example, a wheelbarrow uses a lever and wheel to ease the transport of heavy loads over uneven terrain.

- **Energy Conservation**: Design machines to operate with minimal energy input while maximizing output. For example, a manually operated seed planter can be designed to plant multiple seeds simultaneously, greatly increasing planting speed and reducing human fatigue.

Sustainability

- **Recycled Materials**: Use reclaimed materials like old metal from tools or machines and repurposed wood, which can significantly cut costs and decrease the environmental footprint of new construction.

- **Local Sourcing**: By utilizing materials that are locally available, you can reduce the carbon emissions associated with transporting goods from far away, support the local economy, and ensure that the materials are suitable for the local environment.

Building simple machines for off-grid agriculture and woodworking empowers self-sufficiency and enhances productivity. By applying these principles, off-grid residents can create valuable tools that are both effective and environmentally conscious, perfectly suited for their unique living conditions.

Repairing and Maintaining Equipment

Repair and maintenance are crucial for ensuring the longevity and efficiency of equipment used in off-grid living. Due to the remote nature and potential lack of immediate professional support, having a solid strategy for upkeep and repair is essential. This section covers key aspects of maintaining and repairing off-grid technology and tools.

Setting Up a Maintenance Workshop for Off-Grid Living

Creating an effective maintenance workshop is essential for sustaining the functionality and reliability of tools and equipment in an off-grid setting. Here's a detailed approach to setting up a space that serves as both a repair hub and a storage area for all your maintenance needs.

Tools and Supplies

- **Essential Tools**: Equip your workshop with essential tools that cover a broad range of repair tasks. This should include:

  o **Hand Tools**: Wrenches (adjustable and fixed), screwdrivers (various sizes and heads), pliers (needle nose, locking, and standard), hammers, and a set of Allen wrenches.

  o **Power Tools**: Depending on your access to power, include items like a drill, a circular saw, and a grinder. Battery-operated versions can be particularly useful if power supply is inconsistent.

  o **Specialty Tools**: Depending on the specific needs, tools such as a multimeter for electrical projects, a welding machine, or a pipe bender might be necessary.

- **Supplies**: Stockpile a variety of supplies that will be required for maintenance and repairs:

  o **Consumables**: Oils and greases for lubrication, sandpaper for smoothing surfaces, and sealants for leaks.

  o **Hardware**: Assortments of screws, nails, nuts, bolts, washers, and hinges.

  o **Spare Parts**: Keep on hand spare parts that are specific to your most critical equipment, such as belts for machinery, spare blades for saws, and replacement parts for any small engines.

Organized Workspace

- **Storage Solutions**: Implement storage solutions that keep tools and materials well-organized and protected from the elements. Use toolboxes, pegboards, shelving units, and drawers. Label each storage space to make finding and returning tools straightforward.

- **Work Surfaces**: Designate work surfaces for different types of tasks. Include a sturdy workbench for heavy-duty work and a separate, clean table for more delicate projects like electrical repairs.

- **Safety Measures**: Ensure that your workspace is safe by installing good lighting and fire safety equipment, including a fire extinguisher and first aid kit. Keep all pathways clear to avoid tripping hazards.

- **Environmental Control**: If possible, include some form of environmental control to protect sensitive tools from rust and corrosion. This could be as simple as a dehumidifier or insulation to maintain a stable temperature and humidity level.

Efficiency Tips

- **Tool Maintenance**: Regularly clean and maintain your own tools to set a standard for care and longevity. Oil moving parts, sharpen blades, and check for damages to handles or electrical cords.

- **Workflow Layout**: Arrange your workshop so that you can move efficiently from one task to the next. Place frequently used tools within easy reach, and store heavier equipment near where it will be used to minimize lifting and moving.

Regular Maintenance Procedures for Off-Grid Equipment

Maintaining your tools and machinery is crucial in an off-grid environment, where replacements and professional repairs might not be readily available. Here's a systematic approach to regular maintenance that can help extend the lifespan of your equipment and ensure it remains functional and safe to use.

Inspection

- **Routine Checks**: Establish a regular schedule to inspect all equipment. This could be monthly, quarterly, or semi-annually depending on the equipment's usage frequency and critical nature.

- **What to Look For**: During inspections, check for:

  o **Structural Integrity**: Look for cracks, fractures, or any structural damage in the equipment frames.

  o **Signs of Wear and Tear**: Examine all parts for excessive wear, such as thinning belts, dull blades, or stretched cables.

  o **Leaks and Rust**: Identify any fluid leaks or accumulation of rust, especially in metal components and connections.

  o **Electrical Systems**: For equipment with electrical components, check for frayed wires, loose connections, and signs of short-circuiting.

Cleaning

- **Post-Use Cleaning**: Clean equipment immediately after use to remove dirt, plant residues, or other materials that can interfere with functionality.

- **Deep Cleaning**: Schedule periodic deep cleans that involve dismantling equipment to reach internal components. This is particularly important for complex machinery like engines and hydraulic systems.

- **Proper Techniques and Tools**: Use appropriate cleaning agents and tools such as brushes, cloths, and compressed air to effectively remove debris without damaging the equipment.

Lubrication

- **Lubricant Selection**: Choose the right type of lubricant based on the manufacturer's recommendations and environmental conditions. Some equipment may require biodegradable lubricants if used near water sources or sensitive ecosystems.

- **Application Frequency**: Lubricate all moving parts according to a regular schedule or after cleaning. Pay special attention to high-friction areas and those exposed to high levels of dust and dirt.

- **Monitoring**: After lubrication, monitor the equipment during use to ensure there is no excessive noise or heat, which can indicate inadequate lubrication.

Documentation

- **Maintenance Logs**: Keep detailed records of all maintenance activities, including dates, types of maintenance performed, and any parts replaced or repaired. This helps in tracking the history of each piece of equipment and can be valuable for troubleshooting future issues.

- **Review and Adjust**: Regularly review maintenance logs to identify any recurring issues or trends. Adjust maintenance schedules as necessary based on equipment performance and these findings.

Repair Techniques for Off-Grid Equipment

Maintaining and repairing your equipment in an off-grid setting is essential for continuity and efficiency. Here's a detailed guide to developing effective repair techniques that ensure your machinery and tools remain in top working condition.

Troubleshooting

- **Understanding Equipment**: Familiarize yourself with the manuals and operational guides of your equipment. Understanding how each piece functions will aid in quick identification of issues when they arise.

- **Systematic Approach**: Develop a systematic approach to troubleshooting. Start by checking the most common problems and then move to more complex diagnostics. For example, if a machine fails to start, check the power source first, then the switches, and finally the internal components.

- **Use of Diagnostic Tools**: Employ basic diagnostic tools such as multimeters for electrical equipment, pressure gauges for hydraulic systems, and stethoscopes for detecting unusual noises in engines.

Replacement of Parts

- **Spare Parts Inventory**: Maintain an inventory of essential spare parts based on your equipment's usage and the likelihood of failure. Include items like belts, filters, seals, and bearings.

- **Replacement Skills**: Learn the skills necessary to replace parts. This might include basic mechanical skills such as disassembling and reassembling parts, soldering, or welding.

- **Correct Installation**: Ensure that replacement parts are installed correctly to avoid further damage. Use the correct tools and follow the manufacturer's guidelines for installation.

Resourcefulness

- **Creative Solutions**: Develop the ability to think creatively about repairs. For instance, if a specific gasket is unavailable, you might cut a new one from an old rubber sheet.

- **Improvisation Skills**: Improvise repairs with what you have on hand. For example, if a handle breaks on a tool, fashion a new one from scrap wood or metal.

- **Temporary Fixes**: Learn how to make temporary repairs that can keep equipment operational while you source or wait for replacement parts. For example, using wire to temporarily hold a component in place or applying epoxy putty to seal small leaks.

Learning from Repairs

- **Documentation**: Document each repair, noting what went wrong, how it was fixed, and any changes made to the original setup. This information can be invaluable for future troubleshooting.

- **Feedback Loop**: Use feedback from repaired equipment to improve your maintenance routines. If certain parts frequently fail, consider upgrading the material or checking for underlying issues that may be causing the failures.

Advanced Repair Skills for Off-Grid Living

Developing advanced repair skills is crucial for maintaining and enhancing the longevity and functionality of your tools and machinery, especially in remote or off-grid environments. Here's a breakdown of some essential advanced skills that can greatly benefit anyone living off-grid.

Welding

- **Basic Techniques**: Learn basic welding techniques such as MIG, TIG, and stick welding. Each type has its advantages depending on the type of repair and the materials involved.

- **Safety First**: Understand the safety aspects of welding, including the use of proper personal protective equipment (PPE) like welding helmets, gloves, and protective clothing.

- **Practical Applications**: Use welding to repair broken metal tools, machinery frames, and even fabricate new metal structures. This skill reduces the need to purchase new parts and extends the life of existing equipment.

Electrical Repairs

- **Understanding Electrical Systems**: Gain a fundamental understanding of electrical systems used in off-grid setups, including DC and AC systems, inverters, batteries, and solar panels.

- **Hands-On Skills**: Learn to diagnose and repair common electrical issues, such as replacing faulty wiring, fixing or replacing broken fuses, and securing loose connections. Understanding how to safely handle solar panel connections and battery banks is particularly important.

- **Tool Proficiency**: Become proficient with electrical tools like multimeters, wire strippers, and soldering irons, which are essential for performing accurate and safe repairs.

Fabrication

- **Metal Fabrication**: Acquire skills in cutting, bending, and shaping metal to create or repair parts. Knowledge in using tools like angle grinders, bench presses, and drills is beneficial.

- **Woodworking**: Enhance your ability to work with wood, from crafting replacement handles for tools to building parts for machinery or housing structures.

- **Custom Solutions**: Develop the ability to think innovatively to fabricate solutions when exact replacements are not available. This might include altering an existing part to fit a new purpose or crafting something entirely new from scrap materials.

Advanced Tool Use

- **Precision Tools**: Learn to use precision tools for specific repairs, such as calipers for measuring and alignment tools for ensuring the correct configuration of mechanical systems.

- **Heavy Equipment Operation**: For those with access to more complex machinery, gaining skills in operating heavy equipment can aid in larger construction and repair tasks.

Combining Skills for Complex Repairs

- **Interdisciplinary Approach**: Often, complex repairs require a combination of skills. For example, fixing a broken generator might involve both electrical skills to handle the wiring and mechanical skills to deal with moving parts.

- **Continuous Learning**: Stay updated with the latest technologies and techniques in your area of expertise. Online courses, local workshops, and relevant books can provide new insights and skills.

Networking and Collaboration

- **Community Workshops**: Participate in or organize community workshops where individuals can share knowledge and work together on larger or more complex projects.

- **Professional Guidance**: When possible, consult with professionals to learn from their experience, gaining insights that can be adapted for off-grid living.

Mastering these advanced repair skills not only enhances your self-sufficiency but also empowers you to maintain a higher standard of living while off the grid. These skills ensure that you are well-prepared to handle almost any repair or maintenance issue that might arise, saving time and resources in the long run.

Documentation and Manuals for Effective Equipment Management

Proper documentation and the maintenance of manuals are crucial components in managing and maintaining equipment, especially in environments where professional help may not be readily available. Here's how to effectively organize and utilize documentation and manuals for your equipment:

Keep Manuals

- **Accessibility**: Store the manufacturer's manuals for all equipment in a designated, easily accessible location. This ensures that you can quickly refer to them when needed.

- **Digital Backups**: Where possible, keep digital copies of these manuals. Digital copies can be stored on a computer, a USB drive, or even in cloud storage, making them accessible from various devices and safeguarding against physical damage to paper copies.

- **Organized Filing System**: Organize manuals by equipment type or area of use (e.g., kitchen appliances, farm machinery, power tools) to streamline the search process.

Record Keeping

- **Maintenance Logs**: Maintain detailed logs of all repairs, maintenance activities, and part replacements. Include dates, descriptions of the work performed, and any parts used during the process.

- **Digital Recording**: Utilize digital tools like spreadsheets or maintenance management software to keep your records organized and searchable. This digital approach can be particularly helpful in tracking patterns in equipment wear or recurrent issues.

- **Visual Documentation**: Incorporate photographs or videos of the equipment before and after repairs. Visual records can be extremely useful for troubleshooting future problems or for guiding someone else through a repair process.

Benefits of Effective Documentation

- **Improved Maintenance**: Having immediate access to manuals and a clear history of maintenance activities helps in planning regular maintenance schedules, which can extend the life of equipment.

- **Cost Efficiency**: Documented records help in diagnosing recurring issues quickly, reducing the time and money spent on figuring out problems. This can also help in making informed decisions about whether to repair or replace a piece of equipment.

- **Enhanced Troubleshooting**: Manuals often contain troubleshooting guides that can be invaluable in diagnosing and resolving issues without external help.

- **Knowledge Transfer**: Well-kept records and manuals can assist in training new users or workers, ensuring that they have all the necessary information to operate and maintain the equipment safely and effectively.

Implementing a Documentation Strategy

- **Training**: Train all relevant personnel in the importance of documentation and how to properly use and update the logs and manuals.

- **Regular Reviews**: Schedule regular reviews of the documentation practices and update them as necessary to accommodate new equipment or changes in maintenance procedures.

DIY: Practical Everyday Off-Grid Projects

Engaging in DIY projects is not just a way to solve problems in an off-grid lifestyle; it's a path to greater self-sufficiency and environmental stewardship. Here are several practical, everyday projects that anyone living off-grid can undertake to improve their quality of life, efficiency, and sustainability.

Compost Toilet: DIY Off-Grid Waste Management Solution

Project Overview: A compost toilet is an eco-friendly sanitation system that decomposes human waste into compost, using no water. This type of toilet is ideal for off-grid living, as it helps conserve water and returns nutrients to the earth without contaminating water resources.

Materials Needed:

1. **Two Composting Chambers**: To allow one chamber to compost fully while the other is in use. Construct these from durable, non-corrosive materials like plastic or coated metal.

2. **Toilet Seat**: Can be salvaged from old toilets or purchased new.

3. **Sawdust or Straw**: These carbon-rich materials help absorb moisture, control odor, and balance the carbon-to-nitrogen ratio, which is crucial for effective composting.

4. **Vent Pipe**: Essential for aerating the compost and venting any gases that are produced during the decomposition process.

5. **Urine Diverter**: Diverts urine to a separate container, helping to manage moisture levels in the compost and reduce odors.

Construction Steps:

1. **Build or Assemble the Chambers**: Construct two separate chambers that can be alternated for use. Each should be tightly sealed but accessible for adding waste and removing compost.

2. **Install the Toilet Seat**: Securely mount the toilet seat above the compost chamber, ensuring a tight seal to prevent odors and insects from escaping.

3. **Set Up the Ventilation System**: Install a vent pipe that extends from the top of the compost chamber and vents outside to help control odors and moisture.

4. **Arrange for Urine Diversion**: Install a urine diverter under the toilet seat to separate urine from solid waste. Connect this to a container or a secondary system for urine treatment or use.

Maintenance:

- Regularly add carbon material (sawdust or straw) after each use to facilitate aerobic decomposition and odor control.

- Rotate between the two chambers when one becomes full. Allow the full chamber to compost for at least six months while the other is in use.

- Periodically check the compost for excessive moisture and adjust the amount of carbon additive as necessary.

Benefits:

- **Water Conservation**: Completely eliminates the need for water in waste disposal, saving significant amounts of water annually.

- **Pollution Reduction**: Prevents the contamination of water bodies with nutrients and pathogens typically found in sewage.

- **Soil Enrichment**: Produces nutrient-rich compost that can be used to improve garden soil, supporting healthier plant growth and restoring soil biodiversity.

Safety and Usage Tips:

- Ensure that the compost is fully decomposed before using it in gardens, particularly on food crops, to avoid health risks.

- Regularly inspect and maintain the system to prevent leaks, odors, and other operational issues.

A compost toilet is not only a practical solution for off-grid living but also an active step towards a more sustainable lifestyle, minimizing environmental impact while providing valuable outputs from waste products.

Handmade Furniture: Custom Woodworking Off-Grid

Project Overview: Crafting your own furniture using local wood is a fulfilling project that not only provides functional pieces tailored to your living space but also allows you to engage deeply with the materials and processes of basic carpentry. This project can range from simple tables and chairs to more complex items like beds or shelving units.

Materials Needed:

1. **Wood Logs or Planks**: Choose local wood varieties that are known for their durability and beauty. Hardwoods like oak or maple are excellent for furniture, but softer woods like pine can be easier to work with for beginners.

2. **Basic Carpentry Tools**:

   o **Saw**: For cutting wood to the desired lengths.

   o **Hammer**: For assembling pieces using nails.

   o **Nails or Screws**: Choose based on the type of wood and the design specifications.

   o **Sandpaper**: For smoothing the wood surfaces before finishing.

   o **Measuring Tape and Square**: For accurate measurement and alignment.

Construction Steps:

1. **Design Planning**: Sketch your furniture design, considering the dimensions that best fit your space and needs. Plan for the load-bearing capacity and comfort, especially for chairs and beds.

2. **Wood Preparation**: Cut the wood to your specific measurements, using the saw. Sand all pieces to remove rough edges and splinters, ensuring a smooth finish.

3. **Assembly**:

   o Use the hammer and nails (or screws) to join the pieces according to your design. Ensure joints are strong and well-aligned.

   o For more durable joints, consider learning basic joinery techniques such as dovetails or mortise and tenon.

4. **Finishing**: Apply a protective coating such as varnish, stain, or oil to enhance the wood's natural color and protect it from wear and environmental elements.

Benefits:

- **Cost Efficiency**: Significantly reduces the cost of furnishing your space as compared to purchasing store-bought furniture.

- **Customization**: Allows you to create pieces that perfectly fit and complement your living space.

- **Skill Development**: Enhances your woodworking skills, which can be useful for other projects and maintenance tasks.

Maintenance:

- Regularly check furniture for loose joints or damage, and repair as needed.

- Keep wooden furniture clean and dry to prevent decay. Apply additional coats of protective finish as necessary to maintain its appearance and durability.

Environmental and Personal Value:

- Using locally sourced wood not only supports local businesses but also reduces the environmental impact associated with transporting materials.

- Engaging in the process of making your own furniture fosters a deeper appreciation for craftsmanship and sustainability.

This project is an excellent opportunity for off-griders to personalize their living space while practicing and honing their carpentry skills. Each piece of furniture becomes not only a functional item in the home but also a personal statement of self-sufficiency and environmental responsibility.

Vertical Garden: Optimizing Space for Urban Farming

Project Overview: A vertical garden is an excellent solution for growing herbs, vegetables, and flowers in urban areas or any location where ground space is scarce. By utilizing vertical space, this garden type allows for the cultivation of a variety of plants in a compact area, making it ideal for balconies, patios, or even indoor settings.

Materials Needed:

1. **Hanging Baskets and Recycled Containers**: These can be used to hold soil and plants. Opt for containers that are lightweight and durable.

2. **Wall Frames or Vertical Support Structures**: These will be used to secure the containers. Frames can be made from wood, metal, or any sturdy material capable of supporting the weight of filled containers.

3. **High-Quality Soil**: Suitable for container gardening. It should be well-draining and rich in organic matter.

4. **Seeds or Starter Plants**: Choose a variety of plants that thrive in your local climate and sunlight conditions. Herbs, leafy greens, and small vegetables are ideal for vertical gardening.

5. **Drip Irrigation System**: This system will deliver water directly to the roots, reducing water waste and ensuring that each plant receives adequate moisture.

Construction Steps:

1. **Design Your Layout**: Determine the placement of your vertical garden considering the light requirements of the plants. Ensure that the wall or frame used can withstand environmental elements and the weight of the garden.

2. **Prepare Containers**: Drill drainage holes in the bottom of containers if not already present. Add a layer of pebbles or small stones to enhance drainage before adding soil.

3. **Install Wall Frames**: Securely attach frames or supports to the wall or fence where the garden will be located. Ensure they are level and stable.

4. **Planting**: Fill containers with soil and plant seeds or starter plants. Arrange containers on the frame, spacing them to allow room for growth and adequate air circulation.

5. **Set Up Drip Irrigation**: Install the irrigation system to ensure consistent and easy watering. Route the irrigation lines through the containers and connect to a water source.

Benefits:

- **Space Efficiency**: Makes use of vertical space, allowing for more plants in a smaller area.

- **Aesthetic Enhancement**: Adds a green, lush element to urban environments, which can improve the aesthetic of your living space and contribute to better air quality.

- **Fresh Produce**: Provides a convenient source of fresh vegetables, herbs, and flowers, reducing the need to buy these from the store and promoting healthier eating habits.

Maintenance Tips:

- Regularly check and adjust the drip irrigation system to ensure all plants receive adequate water.

- Monitor plant health and look for signs of pests or disease. Treat issues promptly to prevent spread.

- During peak growing seasons, fertilize the plants appropriately to support growth.

Environmental Impact:

- Vertical gardens can help reduce the urban heat island effect by covering warm surfaces with vegetation.

- Promotes biodiversity in urban areas by providing habitats for pollinators and other wildlife.

This project is a sustainable, productive way to enhance urban living spaces and create a functional garden that yields a variety of benefits while utilizing minimal ground space.

Biomass Briquette Press: Creating Eco-Friendly Fuel

Project Overview: The biomass briquette press is an invaluable tool for converting organic waste materials into solid fuel briquettes that can be used for heating or cooking. This DIY project is especially beneficial for those looking to reduce waste and utilize renewable resources effectively.

Materials Needed:

1. **Metal Plates**: These will form the core of the press where the briquettes are shaped.

2. **Hydraulic Jack**: Used to apply pressure to compact the organic materials into briquettes.

3. **Wooden Frame**: Serves as the structural support for the press, holding the metal plates and hydraulic jack in place.

4. **Organic Materials**: Sawdust, leaves, grass clippings, or any other combustible organic waste available. The finer the particles, the better the compression and cohesion of the briquettes.

Construction Steps:

1. **Build the Frame**: Construct a sturdy frame from the wood to hold the hydraulic jack and metal plates. The frame should be designed to withstand significant force without bending or breaking.

2. **Assemble the Press Mechanism**: Position the hydraulic jack between two metal plates within the frame. Ensure that the bottom plate is fixed securely and the top plate can be moved by the jack.

3. **Prepare the Organic Material**: Dry the organic material thoroughly as moisture can hinder the briquetting process. Once dry, pulverize the material into a fine consistency.

4. **Pressing the Briquettes**: Fill the mold made by the metal plates with the organic material. Use the hydraulic jack to apply pressure, compacting the material into a dense block. The pressure should be maintained for several minutes to ensure the material bonds together.

5. **Extraction and Curing**: Carefully remove the compressed briquette and allow it to dry in a well-ventilated area until it hardens completely.

Benefits:

- **Waste Reduction**: Converts agricultural or yard waste into useful products, thereby reducing the amount of waste sent to landfills.

- **Sustainable Energy**: Provides a renewable source of energy, reducing reliance on non-renewable resources like wood or coal.

- **Cost-Effective**: Utilizes freely available materials, saving money on fuel costs over time.

Usage Tips:

- Ensure briquettes are completely dry before use to avoid smoke and incomplete combustion.

- Store briquettes in a dry place to prevent them from absorbing moisture.

Environmental Impact:

- Burning biomass briquettes releases a considerably lower amount of pollutants compared to coal or wood.

- Helps in managing agricultural waste which can otherwise contribute to environmental pollution.

Creating a biomass briquette press is a practical project that not only aids in managing organic waste but also contributes to sustainable energy practices. By using simple materials and tools, you can create a device that helps turn waste into a valuable resource. This not only conserves natural resources but also offers a renewable and economic alternative for energy production.

Book 5: Food

Food Preservation Without Electricity: Drying, Salting, Fermentation

In off-grid living, food preservation is crucial for extending the shelf life of produce and ensuring a stable food supply throughout the year. Traditional methods such as drying, salting, and fermentation are particularly valuable because they do not require electricity. Here's how to implement these age-old techniques effectively.

Drying: An Essential Food Preservation Technique

Drying is one of the simplest and most ancient methods of food preservation, effectively reducing the moisture content in food to prevent the growth and proliferation of bacteria, yeast, and mold. This process not only preserves the food for longer periods but can also concentrate flavors and nutrients, making dried foods both tasty and nutritious.

Methods of Drying

- **Sun Drying**: This is the most traditional method, ideal for regions with high temperatures and low humidity. Sun drying is cost-effective and energy-efficient, as it utilizes the natural power of the sun. Suitable items for sun drying include fruits like apples, peaches, and apricots, as well as vegetables like tomatoes and peppers.

- **Solar Dehydrators**: These devices enhance the efficiency of sun drying and are particularly useful in areas where the sunlight is less intense or more variable. Solar dehydrators control the air flow and temperature to create the perfect drying conditions, which speeds up the process and can result in a more consistent product.

- **Air Drying**: This method is particularly suitable for herbs, flowers, and mushrooms. These items are best dried in a warm, dry, and well-ventilated area away from direct sunlight. Hanging bunches of herbs or laying items on screens are popular air drying techniques.

Preparation for Drying

- **Slicing**: Cut fruits and vegetables into thin, uniform slices to ensure even drying. This reduces the drying time and prevents some pieces from becoming too dry while others are still moist.

- **Blanching**: For certain vegetables, such as carrots, greens, and broccoli, blanching before drying can stop enzyme activities that degrade the food. This process involves boiling the vegetables for a short time and then plunging them into ice water. Blanching helps to preserve the color, texture, and nutritional content of vegetables during drying.

- **Pre-Treatment**: Some fruits, like apples and pears, may benefit from a pretreatment such as soaking in lemon juice or an ascorbic acid solution to prevent browning and to retain vitamin content.

Storage of Dried Foods

- **Airtight Containers**: Once dried, foods should be stored in airtight containers to protect them from moisture and pests. Glass jars, metal tins with tight lids, or vacuum-sealed bags are excellent options.

- **Cool, Dark Conditions**: Store the containers in a cool, dark place. Excessive heat or light can cause the dried foods to deteriorate more quickly, losing flavor and nutritional value.

- **Check for Moisture**: Regularly check your stored dried foods for signs of moisture or mold. If any moisture is detected, additional drying may be necessary to prevent spoilage.

Advantages of Drying

- **Long Shelf Life**: Properly dried and stored foods can last for months or even years without the need for refrigeration.

- **Nutrient Retention**: Drying preserves important nutrients, making dried fruits and vegetables a nutritious and convenient snack.

- **Versatility**: Dried foods can be used in a variety of culinary applications, from baking and cooking to snacks and garnishes.

Food preservation through drying is an ancient practice that extends the availability of many foods throughout the year. Here is a list of foods that can be effectively preserved by drying, along with the seasons when these foods are typically harvested:

Fruit

- **Apples** - Harvested in the fall. Can be dried in slices or chips.

- **Pears** - Harvested in late summer and fall. Ideal for slice drying.

- **Peaches** - Harvested in summer. Excellent when dried in slices.

- **Apricots** - Harvested in late spring and early summer. Perfect for drying.

- **Plums** - Harvested in summer. Commonly dried and known as prunes.

- **Figs** - Harvested in late summer and fall. Excellent when dried.

- **Bananas** - Available year-round. Popular as dried banana chips.

Vegetables

- **Tomatoes** - Harvested in summer. Can be dried whole, halved, or in pieces to create sun-dried tomatoes.

- **Peppers** - Harvested in summer. Dried and used for spices or preserved whole.

- **Mushrooms** - Available year-round, but the peak is in spring and fall. Ideal for drying.

- **Zucchini** - Harvested in summer. Can be dried in slices.

- **Carrots** - Harvested year-round. Optimal when dried in slices or julienned.

Herbs

- **Basil** - Grows best in summer. Dried for prolonged use in cooking.

- **Mint** - Best harvested in spring and summer. Perfect for drying.

- **Rosemary** - Harvested in spring and fall. Ideal for drying and year-round culinary use.

- **Sage** - Harvested in spring and fall. Commonly dried for use in cooking.

- **Oregano** - Harvested in summer. Often dried to preserve its pungent aroma.

Drying these foods not only allows for their preservation for use during months when they are not available fresh but also offers a way to enjoy their intensified flavors and concentrated nutritional properties. Choosing to dry fruits, vegetables, and herbs during their peak seasons ensures the best flavor and quality of the final product.

Salting: An Essential Method for Food Preservation

Salting, also known as curing, is a time-honored method of food preservation that relies on the use of salt to draw moisture out of the food through osmosis. This reduction in moisture creates conditions that are unfavorable for the growth of microbes, effectively extending the shelf life of various foods.

Methods of Salting

- **Dry Salting**: This traditional technique involves rubbing salt directly onto the food. The salt pulls moisture from the interior to the surface, where it evaporates. Dry salting is often used for meats and fish but can also be applied to certain vegetables.

- **Brining**: Brining is the process of soaking food in a saltwater solution. This method is particularly effective for foods that contain a lot of moisture, which can dilute the effects of dry salting. Brining is commonly used for meats, fish, and some vegetables like cucumbers for pickles.

Applications of Salting

- **Meats and Fish**: Salting is most famously used for preserving various types of meat and fish. Products like bacon, ham, salted fish, and corned beef are all traditionally made using curing techniques that involve salt.

- **Vegetables**: Certain vegetables can also be preserved using salt. Foods such as cabbage for sauerkraut and Korean kimchi are examples where salting plays a crucial role in both preservation and flavor enhancement.

Preparation for Salting

- **Cleaning**: Thoroughly clean all food items before salting to avoid trapping impurities.

- **Cutting and Slicing**: Larger pieces of meat or vegetables may need to be cut or sliced to ensure that the salt can penetrate more deeply and evenly.

- **Salt Selection**: Use non-iodized salt, such as kosher salt or sea salt, for better flavor and effectiveness.

Storage of Salted Foods

- **Cool, Dry Conditions**: Store salted products in a cool, dry place to prevent moisture reabsorption and microbial growth.

- **Airtight Containers**: Use airtight containers to prevent exposure to air, which can lead to spoilage and quality degradation.

- **Monitoring**: Check stored salted foods regularly for signs of spoilage such as mold or an off smell, especially if they are stored for extended periods.

Benefits of Salting

- **Longevity**: Properly salted and stored, food can last several months without the need for refrigeration, which is invaluable in off-grid living situations.

- **Flavor**: Salting not only preserves food but also enhances its flavor, making it a popular technique in culinary traditions worldwide.

- **Cost-Effective**: Salting is a relatively low-cost method requiring minimal equipment, making it accessible for home use.

Health Considerations

- **Sodium Intake**: Be mindful of the increased sodium content in salted foods, which may not be suitable for everyone's dietary needs.

- **Balanced Diet**: Incorporate salted foods into a balanced diet, complementing them with fresh produce and other low-sodium food options.

Salting is a versatile preservation technique that has been used for centuries to safely extend the storage life of foods. By mastering the methods of dry salting and brining, you can ensure a steady supply of preserved foods that are both tasty and nutritious.

Salting Recipes:

1. Traditional American Country Ham

Ingredients:

- 1 fresh ham (leg of pork, about 18-20 lbs)
- 4 cups of non-iodized salt
- 2 cups of brown sugar
- 1 cup of black pepper
- 1/2 cup of pink curing salt

Instructions:

1. Mix all the dry ingredients thoroughly.
2. Rub the mixture all over the ham, ensuring it is well coated.
3. Place the ham in a large clean container and cover it completely with the remaining salt mixture.
4. Store the container in a cool, dry place for 2-3 days per pound (about 36-60 days).
5. Periodically drain any liquid that accumulates and add additional salt mixture.
6. After curing, hang the ham in a cool, dry place to age for another 6 to 12 months before eating.

2. Homemade Salt-Cured Bacon

Ingredients:

- 5 lbs pork belly
- 1/2 cup kosher salt
- 1/4 cup sugar
- 2 tablespoons black pepper
- 1 teaspoon pink curing salt

Instructions:

1. Combine all the ingredients in a bowl.
2. Coat the pork belly evenly with the cure mixture.
3. Place the pork in a zip-top bag, removing as much air as possible.
4. Refrigerate for 7 to 10 days, turning the bag every day.
5. After curing, rinse the pork belly and pat dry.
6. Hang the pork belly in a cool, ventilated area for about 2 weeks to dry.

3. Beef Jerky

Ingredients:

- 2 lbs lean beef (sirloin or flank steak works well)
- 2/3 cup soy sauce
- 2/3 cup Worcestershire sauce
- 2 tablespoons kosher salt
- 1 tablespoon black pepper
- 1 teaspoon onion powder
- 1 teaspoon garlic powder

Instructions:

1. Slice the beef into very thin strips.
2. Mix all other ingredients in a bowl to create a marinade.
3. Submerge the beef strips in the marinade and refrigerate overnight.
4. Remove the beef from the marinade and pat dry.
5. Arrange the strips on a rack and dry using a dehydrator or an oven set to the lowest temperature until completely dry (typically 4-8 hours).

4. Salt-Cured Venison

Ingredients:

- 3 lbs venison, cut into 1/4 inch thick slices

- 1/2 cup kosher salt
- 1/4 cup sugar
- 1 tablespoon cracked black pepper

Instructions:

1. Combine salt, sugar, and pepper in a mixing bowl.
2. Thoroughly rub the mixture into each slice of venison.
3. Stack the venison in a container, cover, and refrigerate for 3 days.
4. Rinse the salt off with cold water and pat the venison dry.
5. Dry the venison slices in a dehydrator at 140°F until fully dehydrated (about 4-6 hours).

5. Pioneer Salt Pork

Ingredients:

- 5 lbs pork belly or fatback
- 2 cups kosher salt
- 1/2 cup brown sugar

Instructions:

1. Mix the salt and sugar in a large bowl.
2. Coat the pork pieces completely with the salt mixture.
3. Place the pork in a ceramic crock or wooden barrel and cover it with any remaining salt mixture.
4. Store in a cool, dry place for at least one month before use.

6. Fisherman's Salted Cod

Ingredients:

- 4 lbs fresh cod fillets
- 3 cups kosher salt

Instructions:

1. Sprinkle a layer of salt on the bottom of a container.
2. Place a layer of cod fillets over the salt, then cover with another layer of salt.
3. Repeat layering until all the cod is covered in salt.
4. Cover the container and refrigerate for at least 3 weeks.
5. To use, soak the cod in water for 24 hours, changing the water several times to desalinate.

7. Salt-Preserved Citrus Peels

Ingredients:

- Peels from 6 oranges or lemons
- 1 cup kosher salt

Instructions:

1. Scrape any remaining fruit from the peels and cut them into strips.
2. Toss the peels with salt in a bowl.
3. Pack the salted peels into a jar, pressing them tightly.

4. Seal the jar and let sit at room temperature for 3 weeks.
5. Rinse the peels before use in cooking or baking.

8. Applewood Smoked Salt Trout

Ingredients:

- 4 trout fillets
- 1 cup kosher salt
- Wood chips (applewood preferred)

Instructions:

1. Rub the trout fillets thoroughly with kosher salt.
2. Place the fillets in a container, cover, and refrigerate for 12 hours.
3. Rinse the fillets and pat dry.
4. Smoke over applewood chips at 225°F for 4 hours.

9. Salt-Cured Turkey Jerky

Ingredients:

- 2 lbs turkey breast, thinly sliced
- 2/3 cup kosher salt
- 1/2 cup soy sauce
- 2 tablespoons Worce0stershire sauce
- 1 tablespoon smoked paprika
- 1 teaspoon garlic powder

Instructions:

1. Combine all ingredients except turkey to make a marinade.
2. Submerge the turkey slices in the marinade and refrigerate overnight.
3. Remove the turkey from the marinade, rinse lightly, and pat dry.
4. Arrange on a dehydrator tray and dry at 160°F until fully dried (about 6-8 hours).

Fermentation: Preserving Food Through Natural Processes

Fermentation is a traditional method of food preservation that leverages natural bacteria and yeasts to convert sugars into acids, gases, or alcohol. This not only preserves the food but also enhances its flavors and nutritional benefits, often introducing beneficial probiotics that are good for digestion. Here's a guide on how to get started with fermentation, focusing on simple recipes that are easy to try at home.

Methods of Fermentation

- **Simple Recipes to Start**: Begin with basic fermented foods that are straightforward and require minimal special equipment. Popular options include:
    - **Sauerkraut**: Fermented cabbage with a tangy flavor.
    - **Kimchi**: A Korean dish made from fermented vegetables and a variety of seasonings.
    - **Sourdough Starters**: Cultures of flour and water that act as natural leavening agents for baking bread.

Required Equipment

- **Clean Jars**: Use glass jars that can be sealed tightly but allow gases produced during fermentation to escape.
- **Weights**: To keep vegetables submerged below the surface of the liquid, ensuring an anaerobic environment necessary for proper fermentation.
- **Cloths and Bands**: To cover the jars, allowing air to flow while keeping contaminants out.

Preparation for Fermentation

1. **Sanitization**: Thoroughly clean all equipment and your hands to prevent the introduction of harmful microbes. Sterilize jars and weights by boiling them in water for at least 10 minutes.

2. **Cutting and Slicing**: Chop or slice your produce into consistent pieces. This ensures even fermentation and better flavor distribution. For sauerkrait, finely shred the cabbage; for kimchi, chop vegetables into bite-sized pieces.

3. **Salting**: Many fermented foods require salting or creating a brine solution. This helps to draw out water from the ingredients, creating an ideal environment for fermentation.

Fermentation Process

1. **Mixing Ingredients**: Combine your prepared vegetables with salt, and any other seasonings (such as garlic, peppercorns, or caraway seeds for sauerkraut; ginger, garlic, and chili peppers for kimchi).

2. **Packing**: Tightly pack the mixture into your sterilized jars, pressing down until the liquid (brine) covers the vegetables. The brine is crucial as it prevents exposure to air, which can lead to spoilage.

3. **Weighing Down**: Place a clean weight on top of the vegetables to keep them submerged.

4. **Covering**: Cover the jar with a cloth and secure it with a rubber band.

5. **Fermentation Time**: Allow the jars to sit at room temperature, out of direct sunlight. The fermentation time can vary depending on the recipe and your taste preferences, typically ranging from a few days to several weeks.

6. **Checking**: Check your ferment every few days to ensure that the vegetables remain submerged and to remove any scum that may form on the surface.

Storage of Fermented Foods

- **Cool Temperature Storage**: Once the desired level of fermentation is reached, close the jars with lids and move them to a refrigerator or a cool cellar. This slows down the fermentation process and preserves the food for longer periods.

- **Shelf Life**: Properly stored, fermented foods can last for several months. The cold environment halts the fermentation, maintaining the taste and benefits of the food.

Benefits of Fermentation

- **Enhanced Nutritional Value**: Fermentation can increase essential nutrients, make foods easier to digest, and introduce beneficial probiotics.

- **Flavor Development**: Fermentation can transform the basic ingredients into something complex and flavorful, adding depth to meals.

- **Food Preservation**: By creating an acidic environment that is inhospitable to harmful bacteria, fermentation naturally preserves the food.

Fermentation is both an art and a science, allowing for experimentation and variety in what you can create. Starting with simple projects like sauerkraut or kimchi will build your confidence and skills in fermentation, leading to more complex ferments and flavors in your culinary repertoire.

Fermentation Recipes:

1. Classic Sauerkraut

Ingredients:

- 1 medium head of cabbage
- 1 tablespoon kosher salt
- Optional: 1 teaspoon caraway seeds

Instructions:

1. Thinly slice the cabbage.
2. In a large bowl, mix the cabbage with salt (and caraway seeds, if using). Massage the salt into the cabbage until it starts to release liquid.
3. Pack the cabbage tightly into a clean jar, pressing down until the liquid rises above the cabbage.
4. Place a weight on top to keep the cabbage submerged. Cover with a cloth and secure with a rubber band.
5. Allow it to ferment at room temperature for at least 2 weeks. Check occasionally, pressing down the cabbage if it rises above the liquid.
6. Once fermented, seal the jar and store in the refrigerator.

2. New York-Style Sour Pickles

Ingredients:

- 2 pounds cucumbers (small to medium size)
- 4 cups water
- 2 tablespoons kosher salt
- 4 cloves garlic, peeled and smashed
- A few sprigs of fresh dill
- 1 tablespoon whole mustard seeds
- 1 teaspoon black peppercorns

Instructions:

1. Combine water and salt to create a brine, stirring until salt dissolves.
2. Place cucumbers in a large jar. Add garlic, dill, mustard seeds, and peppercorns.
3. Pour the brine over the cucumbers, ensuring they are completely submerged.
4. Place a weight over the cucumbers to keep them submerged. Cover the jar with a cloth and secure with a rubber band.
5. Allow to ferment at room temperature for about 1 week. Taste and if satisfied, seal the jar and refrigerate.

3. Kimchi

Ingredients:

- 1 head Napa cabbage, cut into chunks
- 1/4 cup sea salt
- Water, enough to cover cabbage
- 1 tablespoon grated ginger

- 4 cloves garlic, minced
- 1 teaspoon sugar
- 3 tablespoons water
- 1-5 tablespoons Korean red pepper flakes (adjust to taste)
- 8 ounces Korean radish or daikon, peeled and grated
- 4 green onions, chopped

Instructions:

1. Salt the cabbage and cover with water. Let sit for 1-2 hours.
2. Rinse the cabbage under cold water 3 times and drain in a colander for 15 minutes.
3. Prepare the paste by mixing ginger, garlic, sugar, water, and red pepper flakes into a smooth paste.
4. Combine the cabbage with radish, green onions, and paste.
5. Pack the mixture into a clean jar, pressing down until the brine rises above the vegetables.
6. Place a weight over the vegetables to keep them submerged. Cover the jar with a cloth and secure with a rubber band.
7. Allow to ferment at room temperature for 5 days, then refrigerate.

4. Apple Cider Vinegar

Ingredients:

- Apple scraps (cores and peels)
- 2 tablespoons sugar
- Water to cover

Instructions:

1. Place apple scraps in a large jar and cover with water.
2. Add sugar and stir until dissolved.
3. Cover the jar with a paper towel or cloth and secure with a rubber band.
4. Let sit in a dark, room temperature spot for about 3-4 weeks.
5. Strain out the solids, and then let the liquid ferment for another 4 weeks, covered with a cloth.
6. Store the vinegar in sealed bottles.

5. Kombucha

Ingredients:

- 1 kombucha SCOBY
- 1 gallon filtered water
- 1 cup sugar
- 8 bags black tea (or 2 tablespoons loose tea)

Instructions:

1. Boil water, dissolve sugar, and steep tea until water cools to room temperature.
2. Remove tea bags and pour the sweet tea into a large jar.
3. Add the SCOBY and cover with a cloth, securing with a rubber band.
4. Allow to ferment at room temperature for 7-10 days.
5. Taste the kombucha; if it's to your liking, bottle it, leaving room for carbonation.
6. For flavored kombucha, add fruits or herbs before bottling.
7. Store bottled kombucha at room temperature for 2-3 days for carbonation, then refrigerate.

6. Cortido (Latin American Sauerkraut)

Ingredients:

- 1 medium head cabbage, shredded
- 1 carrot, grated
- 1 onion, thinly sliced
- 1/2 cup apple cider vinegar
- 1 tablespoon kosher salt
- 1 teaspoon dried oregano
- 1/2 teaspoon red pepper flakes (optional)

Instructions:

1. Combine cabbage, carrot, and onion in a large bowl. Sprinkle with salt and massage into the vegetables to release their juices.
2. Stir in apple cider vinegar, oregano, and red pepper flakes.
3. Pack the mixture tightly into a clean jar, pressing down until covered by its own liquid.
4. Place a weight to keep everything submerged. Cover the jar with a cloth and secure with a rubber band.
5. Allow to ferment at room temperature for about 5 days. Check daily to ensure vegetables are submerged, pressing down if necessary.
6. Taste the cortido after 5 days; if it's to your liking, place a lid on the jar and store in the refrigerator.

7. Beet Kvass

Ingredients:

- 3 medium beets, peeled and chopped coarsely
- 1/4 cup whey or additional tablespoon salt (if dairy-free)
- 1 teaspoon sea salt
- Filtered water to cover

Instructions:

1. Place the chopped beets in a quart-sized mason jar.
2. Add whey and salt, then fill the jar with filtered water, leaving about an inch of headspace at the top.
3. Stir to mix everything thoroughly.
4. Cover the jar with a cloth and secure with a rubber band.
5. Allow to sit at room temperature for 2-7 days, depending on the desired sourness.
6. Once fermented, strain the liquid into another container, refrigerate, and drink as a probiotic-rich tonic.

8. Fermented Hot Sauce

Ingredients:

- 1 pound fresh chili peppers, chopped
- 4 cloves garlic, peeled
- 2 tablespoons kosher salt
- 2 teaspoons sugar
- Water to cover

Instructions:

1. Combine all ingredients in a blender, adding just enough water to make a thick paste.
2. Transfer to a jar, leaving some space at the top.
3. Cover with a cloth and secure with a rubber band.
4. Let the mixture ferment at room temperature for 1-2 weeks. Stir daily to prevent mold.
5. Once fermented, puree again if desired for a smoother sauce. Store in the refrigerator.

9. Garlic Dill Green Beans

Ingredients:

- 1 pound fresh green beans, trimmed
- 3 cloves garlic, minced
- 2 tablespoons dill seeds
- 2 tablespoons sea salt
- 4 cups water

Instructions:

1. Combine water and salt to create a brine, ensuring the salt is fully dissolved.
2. Place green beans in a clean jar, and add minced garlic and dill seeds.
3. Pour the brine over the green beans, ensuring they are completely submerged.
4. Place a weight over the green beans to keep them submerged. Cover the jar with a cloth and secure with a rubber band.
5. Allow to ferment at room temperature for about 5-7 days. Check regularly to ensure the beans remain submerged.
6. Once fermentation is complete, replace the cloth with a lid and store in the refrigerator.

Food Preservation Techniques: Water Bath Canning

Canning is a method of preserving food in which the food contents are processed and sealed in an airtight container. Canning provides a shelf life typically ranging from one to five years, although under specific circumstances it can be much longer. A freeze-dried canned product, for example, can last as long as 25 or 30 years in an edible state. Water bath canning is a process of heat processing your jars of food in boiling water, suitable for high-acid foods like fruits, tomatoes, jams, jellies, and pickles.

Preparation for Canning

Canning is a rewarding way to preserve food at its peak freshness and enjoy the fruits of your harvest or seasonal market finds all year round. Proper preparation is key to ensuring the safety and quality of your canned goods. Here's a step-by-step guide to getting ready for canning, covering everything from gathering your materials to preparing your recipes.

Ingredients and Equipment

1. Gather Your Materials:

- **Canning Jars:** Choose the appropriate size for your project, whether pints, quarts, or half-gallons. Ensure they are free of nicks, cracks, or rough edges that could prevent sealing or cause breakage.

- **Lids and Rings:** Use new, unused canning lids each time to ensure a safe seal. Rings can be reused as long as they are not rusty or bent.

- **Canning Pot:** A large pot with a canning rack is essential to safely process jars in boiling water. Ensure the pot is tall enough to allow at least one inch of water to boil over the tops of the jars.

- **Jar Lifter:** A specially designed tool that allows you to safely remove hot jars from boiling water.

- **Funnel:** A wide-mouth funnel helps in transferring food into jars without mess and keeps the jar rims clean.

- **Clean Cloths:** Have clean cloths or paper towels on hand to wipe down jar rims and clean up spills.

2. Sterilization:

- **Jars:** Wash all jars in hot, soapy water, then rinse well. To sterilize, boil them in water for at least 10 minutes or heat them in a preheated oven at 275°F (135°C) for at least 20 minutes.

- **Lids and Rings:** Lids should also be washed in hot, soapy water and can be simmered (not boiled) for a few minutes to sterilize. Ensure the sealing compound on the lids is softened for a secure fit.

Prepare Your Recipe

1. Choose Your Recipe: Select a recipe from a reliable source. Recipes designed for canning will consider the necessary acid levels and processing times to ensure safety from foodborne pathogens like botulism.

2. Preparing Ingredients:

- **Fruits and Vegetables:** Wash thoroughly. Peel, chop, slice, or prepare as specified in your recipe. For fruits, you might need to add lemon juice or ascorbic acid to prevent browning.

- **Brines, Syrups, and Seasonings:** Prepare according to your recipe. For pickles, a vinegar brine is commonly used. For fruits, a light syrup might be preferred, which involves boiling water and sugar.

- **Cooking Down:** Some recipes may require you to cook your fruits or vegetables before canning. This could involve simmering berries for jams or cooking tomatoes down into sauce.

Water Bath Canning Process

Water bath canning is a popular method for preserving high-acid foods such as fruits, tomatoes, jams, jellies, and pickles. This method uses boiling water to safely process canned food, ensuring it's preserved for long-term storage without refrigeration. Here's how to perform the water bath canning process from start to finish:

Filling the Jars

1. **Prepare Your Jars**: Ensure your jars are clean and hot before you begin filling them. You can keep them warm in simmering water or a heated dishwasher.

2. **Use a Funnel**: Place a wide-mouth funnel on the jar and carefully fill the jars with your prepared food, following your specific recipe. Be sure to leave the recommended headspace, usually about 1/2 inch, to allow for expansion during heating.

Removing Air Bubbles

1. **Slide a Spatula**: After filling, take a non-metallic spatula or a bubble remover tool and gently slide it between the food and the jar. Move it around the edges to release any trapped air bubbles. This step ensures that your food settles properly and minimizes the risk of spoilage.

Wiping Rims and Sealing

1. **Clean the Rims**: Wipe the rims of the jars with a clean, damp cloth or paper towel. This removes any food particles or residues that might prevent a proper seal.

2. **Apply Lids and Bands**: Place the cleaned lids on the jars, ensuring the sealing compound is in contact with the jar rim. Screw on the bands until they are fingertight. It's important not to overtighten, as air must be able to escape during processing.

Processing

1. **Prepare the Canner**: Place the rack in the bottom of the canning pot and fill the pot with enough water to cover the jars by at least one inch. The water should be warm but not boiling when you place the jars in the pot.

2. **Boil the Water**: Bring the water to a full rolling boil. Once boiling, lower the jars into the water using a jar lifter, ensuring they don't touch each other. Cover the pot and continue to maintain a steady boil for the entire processing time recommended in your recipe.

Cooling Jars

1. **Turn Off Heat**: Once the processing time is complete, turn off the heat and let the jars sit in the water for 5 minutes to stabilize.

2. **Remove Jars**: Using a jar lifter, carefully remove the jars from the pot and place them on a towel or cooling rack, ensuring they are not in a drafty area.

3. **Cool Completely**: Allow the jars to cool for 12 to 24 hours. Do not disturb them during cooling.

Checking Seals

1. **Test the Seals**: After the jars have cooled, check the seal by pressing the center of the lid. If the lid does not pop up and down, the jar is sealed.

2. **Reprocess if Necessary**: If a lid fails to seal, you can either store the jar in the refrigerator and use the contents soon, or you can reprocess the contents. If reprocessing, check the jar, lid, and contents for any signs of spoilage, replace the lid, and process again as before.

Storage

1. **Label and Store**: Once all jars have been checked and are properly sealed, label them with the contents and the date. Store in a cool, dark place. Properly canned goods can last for up to a year.

Following these steps ensures that your canned foods are safely processed, allowing you to enjoy your homemade preserves throughout the year. This method is not only practical but also a rewarding way to capture the flavors of the season.

Benefits of Water Bath Canning

Water bath canning is a time-tested method of food preservation that offers several key benefits, making it a favorite among home gardeners and cooking enthusiasts. Here are some of the main advantages of using this method:

Longevity

- **Extended Shelf Life**: One of the most significant benefits of water bath canning is the ability to extend the shelf life of fresh foods. Properly canned and sealed foods can last for up to a year or more without refrigeration. This is especially valuable for seasonal fruits and vegetables, allowing you to enjoy summer's peak flavors even in the depth of winter.

Nutritional Preservation

- **Retention of Nutrients**: While some nutrients may be lost during the cooking process, canning generally preserves the majority of vitamins and minerals. Water-soluble vitamins like Vitamin C and B vitamins are better retained during canning compared to prolonged storage of fresh produce. The quick heat process and sealed environment help minimize nutrient loss, ensuring that the canned foods are not only tasty but also nutritious.

Enhanced Flavor

- **Flavor Development**: Canning can enhance the natural flavors of food. Fruits and vegetables canned in syrups or brines absorb additional flavors over time, often resulting in a more intense and complex flavor profile compared to their fresh counterparts. This makes canned goods particularly delightful in culinary applications where enhanced flavors can elevate a dish.

Safety

- **Food Safety**: Proper canning techniques destroy microorganisms that can cause food spoilage and foodborne illnesses. The high heat of the boiling water bath ensures that any potential bacterial growth is halted, providing a safe, sterile environment for food storage.

Convenience

- **Ready-to-Use Convenience**: Canned foods are fully cooked and ready to use, which can save a significant amount of preparation time in the kitchen. Whether you're making a quick meal or planning an elaborate dish, having jars of canned tomatoes, fruits, or pickles on hand means you're always ready to go.

Cost-Effective

- **Economical**: Water bath canning offers an economical way to preserve large quantities of food. It utilizes equipment that is relatively inexpensive and reusable. By canning your own food, you can also reduce food waste and control what goes into your diet, avoiding the additives and preservatives found in many commercially canned products.

Eco-Friendly

- **Sustainability**: By preserving your own food, you reduce the need for commercially processed foods, which often involve higher energy use and more packaging. Canning at home uses less energy per batch and results in reusable and recyclable materials, which is a more sustainable option for the environmentally conscious.

These benefits highlight why water bath canning is such a popular method for preserving food. It not only enhances the safety, taste, and nutritional value of foods but also contributes to a more sustainable and economical way of managing food resources.

Safety Tips for Water Bath Canning

Water bath canning is a fantastic way to preserve food, but it must be done correctly to ensure safety. Here are some crucial safety tips to keep in mind:

Acid Levels

- **Correct Acidity**: It's imperative to use recipes that are specifically designed for water bath canning. These recipes ensure that the acidity levels in the food are high enough to safely can using this method. Foods like fruits, tomatoes, pickles, and jams generally have sufficient acidity, either naturally or through the addition of lemon juice or vinegar.

- **Use Reliable Sources**: Only use recipes from reputable sources, such as extension services or established canning books. The correct acid level is crucial for preventing the growth of Clostridium botulinum bacteria, which cause botulism, a potentially fatal illness.

Lid Usage

- **Single Use for Lids**: Canning lids are designed to seal only once. Reusing lids can result in a poor seal, leading to food spoilage or contamination. Always use new lids each time you can to ensure a safe and secure seal.

- **Inspect Lids**: Before using, inspect lids for any dents, rust, or imperfections that might prevent them from sealing correctly.

General Equipment Safety

- **Sterilization**: Thoroughly sterilize all canning jars, lids, and other equipment before use. This can typically be done by boiling them for 10 minutes or washing in a dishwasher with a sterilize setting.

- **Avoid Cross Contamination**: Keep your workspace clean and manage your tools in a way that prevents the introduction of bacteria. Use separate spoons, ladles, and funnels to handle different products, especially when dealing with multiple batches.

Process Control

- **Heating and Cooling**: Ensure that jars are filled while the food and jars are hot. Place them in boiling water in the canner, making sure water covers the jars by at least one inch. Follow exact processing times as per the recipe to ensure all pathogens are destroyed.

- **Cooling Jars**: Allow jars to cool naturally in a draft-free area after processing. Do not attempt to hasten the cooling process as this can lead to unsealed jars or jar breakage.

Monitoring and Storage

- **Check Seals**: After the jars have cooled for 12 to 24 hours, check the seals by pressing down on the center of the lid. If the lid pops back, it is not sealed and the food should not be stored. You can either reprocess the food within 24 hours or refrigerate it and use it immediately.

- **Proper Storage**: Store sealed jars in a cool, dark place to prevent spoilage and fading of colors and flavors.

By adhering to these safety guidelines, you can ensure that your water bath canning process is both safe and effective, providing you with delicious, preserved foods that are enjoyable throughout the year.

Water Bath Canning **Recipes:**

1. Classic Apple Butter

- **Ingredients**: Apples, sugar, cinnamon, cloves, allspice.
- **Process**: Cook apples down into a thick butter, season, and can.

2. Strawberry Jam

- **Ingredients**: Fresh strawberries, sugar, pectin, lemon juice.
- **Process**: Mash strawberries, combine with other ingredients, boil, and can.

3. Dill Pickles

- **Ingredients**: Cucumbers, dill, garlic, vinegar, water, pickling salt.
- **Process**: Pack jars with cucumbers and dill, pour boiling vinegar brine over, and process.

4. Tomato Sauce

- **Ingredients**: Tomatoes, onions, garlic, olive oil, herbs.
- **Process**: Cook down to a sauce, season, and can.

5. Peach Preserves

- **Ingredients**: Peaches, sugar, lemon juice.
- **Process**: Chop peaches, cook with sugar and lemon, can.

6. Blueberry Syrup

- **Ingredients**: Blueberries, sugar, water, lemon juice.
- **Process**: Cook berries with sugar and water, strain, can the syrup.

7. Corn Relish

- **Ingredients**: Corn, onion, red bell pepper, celery, vinegar, sugar, mustard seed, celery seed.
- **Process**: Combine ingredients, boil, and can.

8. Bread and Butter Pickles

- **Ingredients**: Thinly sliced cucumbers and onions, vinegar, sugar, mustard seeds, celery seeds, turmeric.
- **Process**: Soak cucumbers and onions in ice water, drain, cover with hot vinegar solution, and can.

9. Apple Sauce

- **Ingredients**: Apples, water, sugar, lemon juice (optional).
- **Process**: Cook apples until soft, puree, can.

10. Spicy Salsa

- **Ingredients**: Tomatoes, onions, jalapeños, garlic, cilantro, lime juice, vinegar.
- **Process**: Chop and combine all ingredients, cook briefly, can.

Raspberry Jam

- **Ingredients**: Fresh raspberries, sugar, lemon juice, pectin.
- **Process**: Crush raspberries, mix with sugar and pectin, boil until set, and can.

12. Spiced Pear Butter

- **Ingredients**: Pears, sugar, cinnamon, nutmeg, allspice, lemon juice.
- **Process**: Cook pears until soft, puree, add spices and lemon, cook until thickened, and can.

13. Pickled Green Beans

- **Ingredients**: Green beans, vinegar, water, garlic, dill, red pepper flakes.
- **Process**: Blanch green beans, pack into jars with spices, cover with hot vinegar solution, and process.

14. Marinara Sauce

- **Ingredients**: Tomatoes, olive oil, onions, garlic, basil, oregano.
- **Process**: Sauté onions and garlic, add tomatoes and herbs, simmer, and can.

15. Blackberry Syrup

- **Ingredients**: Blackberries, sugar, water, lemon juice.
- **Process**: Simmer blackberries with sugar and water, strain, can the syrup.

16. Hot Pepper Jelly

- **Ingredients**: Bell peppers, hot peppers, sugar, vinegar, pectin.
- **Process**: Puree peppers, mix with sugar and vinegar, boil with pectin, and can.

17. Cranberry Sauce

- **Ingredients**: Cranberries, sugar, water, orange zest.
- **Process**: Combine all ingredients, cook until cranberries burst, and can.

18. Chili Sauce

- **Ingredients**: Tomatoes, onions, bell peppers, vinegar, sugar, spices.
- **Process**: Chop vegetables, combine with vinegar and spices, simmer, and can.

19. Pickled Beets

- **Ingredients**: Beets, vinegar, sugar, water, cloves, cinnamon.
- **Process**: Cook beets, peel and slice, pack into jars, cover with hot vinegar solution, and process.

20. Apricot Preserves

- **Ingredients**: Apricots, sugar, lemon juice.
- **Process**: Halve and pit apricots, cook with sugar and lemon until thickened, and can.

Book 6: Renewable Energy

Introduction to Renewable Energies

Renewable energy, often referred to as clean energy, comes from natural sources or processes that are constantly replenished. Unlike fossil fuels, which are finite and emit harmful pollutants, renewable energy sources are sustainable and environmentally friendly. They play a crucial role in reducing greenhouse gas emissions, combating climate change, and providing a diverse range of energy solutions. This introduction aims to explore the most common types of renewable energy, their benefits, and their applications.

Types of Renewable Energy

1. **Solar Energy**

    o   **Description**: Solar power is derived from the Sun's rays and can be converted into electricity or heat using panels and mirrors. Photovoltaic (PV) panels convert sunlight directly into electricity, while solar thermal technology uses sunlight to heat water or air for residential, commercial, or industrial use.

    o   **Applications**: Residential homes, commercial buildings, solar farms, off-grid applications.

2. **Wind Energy**

    o   **Description**: Wind energy harnesses the power of the wind through turbines that convert kinetic energy into electricity. Wind farms can be located onshore (land) or offshore (over water).

    o   **Applications**: Large-scale power generation, distributed power for remote locations, supplementing existing grids.

Solar Energy: Harnessing the Power of the Sun

Solar energy, one of the most prominent and rapidly growing renewable energy sources, harnesses the sun's abundant energy to provide electricity and heat. It utilizes two primary technologies: photovoltaic (PV) panels and solar thermal systems, each playing a crucial role in reducing reliance on fossil fuels.

Description of Solar Energy Technologies

1. **Photovoltaic (PV) Panels**

   o **Mechanism**: PV panels contain cells made from semiconductor materials, typically silicon, which absorb sunlight. When the sun's photons hit these cells, they generate an electric field across the layers, causing electricity to flow. This process of converting light (photons) to electricity (voltage) is known as the photovoltaic effect.

   o **Use**: The electricity generated is either used directly, stored in batteries for later use, or fed into the grid.

2. **Solar Thermal Technology**

   o **Mechanism**: Unlike PV panels that convert sunlight directly into electricity, solar thermal systems use sunlight to produce heat. These systems include various setups such as flat-plate collectors, evacuated tubes, and parabolic mirrors. They absorb solar radiation to heat fluids (water or an antifreeze mixture), which can then be used to provide space heating, hot water, or even power turbines for electricity generation.

o **Use**: The heat generated can be used immediately or stored in thermal storage systems for later use.

Applications of Solar Energy

1. **Residential Homes**

   o Solar panels can be installed on rooftops to generate electricity, significantly reducing household energy bills and environmental footprint.

   o Solar water heaters can provide an efficient, cost-effective method for generating hot water for domestic use.

2. **Solar Farms**

   o Large-scale solar farms, consisting of thousands of solar panels spread over extensive areas, generate massive amounts of electricity. This electricity is usually fed into the national grid to serve multiple consumers.

   o These farms are typically located in areas with high solar radiation and are crucial in reducing a country's carbon emissions.

3. **Off-Grid Applications**

   o Solar energy is vital in remote areas where traditional electricity access is unreliable or non-existent.

   o Solar panels can power everything from small handheld devices to entire communities, providing a lifeline in off-grid and impoverished areas.

Advantages of Solar Energy

- **Sustainability**: Solar energy is plentiful and sustainable, offering a virtually inexhaustible supply of power without the adverse environmental impacts associated with fossil fuel extraction and use.

- **Decreasing Costs**: As technology advances and production scales, the cost of solar panels and related components continues to drop, making it more accessible to a wider range of people and businesses.

- **Flexibility and Scalability**: Solar installations can be customized to fit specific needs, whether small residential setups or large, multi-megawatt commercial projects.

- **Low Maintenance**: Once installed, solar panels require little maintenance and provide a reliable power supply for many years, typically with warranties lasting 25 years or more.

Solar energy not only promotes environmental stewardhood but also provides economic benefits by creating jobs in the growing tech industry and reducing energy costs. As solar technology continues to evolve, its role in global energy markets is set to increase, making it a cornerstone of future energy systems.

Installing a DIY Solar System

Installing a DIY solar system can be an empowering way to achieve energy independence and reduce your carbon footprint. This project involves planning, purchasing, and assembling the components needed to generate electricity from the sun. Here's a comprehensive guide to help you through the process.

Assessment of Energy Needs

Before installing a solar system, a thorough assessment of your energy needs is essential. This evaluation helps determine the appropriate size and specifications for your system to ensure it meets your household's electricity requirements effectively. Here's how to conduct this assessment:

Calculate Energy Consumption

- **Review Electricity Bills**: Start by examining your past electricity bills, typically for the last year. This will give you an idea of your total energy usage patterns and peak consumption periods. It's important to note the kilowatt-hours (kWh) consumed each month.

- **Identify High Energy Use Appliances**: List down appliances that consume the most power, such as HVAC systems, water heaters, and refrigerators. Knowing which appliances use the most energy can help you plan a system that either focuses on total household supply or targeted high-demand areas.

Evaluate Sunlight Exposure

- **Site Assessment**: Check the proposed location for solar panel installation—commonly the roof. Ensure the site is free from shade for the majority of the day. Obstacles such as trees, neighboring buildings, or architectural features can significantly impact the effectiveness of your solar panels.

- **Sunlight Hours**: Determine the average daily and seasonal sunlight hours your location receives. More sunlight means more potential electricity generation. Solar maps available online or data from local weather services can assist in this evaluation.

- **Panel Orientation and Tilt**: The orientation (azimuth) and tilt of your solar panels affect their capacity to capture sunlight. In the northern hemisphere, solar panels should ideally face south at an angle that equals the latitude of your location to maximize exposure.

Additional Considerations

- **Future Energy Needs**: Consider any planned changes or additions to your household that might increase energy consumption, such as purchasing an electric vehicle or adding new electrical appliances.

- **Energy Efficiency Improvements**: Prior to sizing your system, it might be beneficial to implement energy efficiency measures. Upgrading to energy-efficient appliances, adding insulation, and using LED lighting can reduce overall energy demand, thereby decreasing the size and cost of the solar system required.

Conducting a detailed assessment of your energy needs and sunlight exposure sets a solid foundation for designing an efficient solar power system tailored to your specific requirements. This step is crucial in ensuring that you invest in a solar solution that is both cost-effective and capable of meeting your energy needs.

System Design and Component Selection

When designing a DIY solar system, it's crucial to consider the type of system that best suits your needs, the components required, and the specifications of those components. Here's how to approach each aspect:

Type of System

1. **Grid-Tied System**:

    o **Description**: Connects with the local utility grid, allowing you to feed excess generated power back to the grid through a system known as net metering.

    o **Benefits**: You can receive credits from your utility company for the excess power you supply, reducing your utility bills. Grid-tied systems do not typically require batteries, which lowers the initial cost.

    o **Considerations**: Grid-tied systems will not provide power during a blackout since they must automatically disconnect for safety reasons unless equipped with a grid-tie inverter with backup functionality.

2. **Off-Grid System**:

    o **Description**: Completely independent from the utility grid. This system requires a battery bank to store energy for use when sunlight is not available.

    o **Benefits**: Ideal for remote locations where connecting to the grid is too expensive or impossible. Provides complete energy independence.

    o **Considerations**: Requires careful planning to ensure battery capacity is sufficient to meet energy needs during non-sunny periods. Generally, more expensive due to additional components like batteries and a more robust charging system.

Components Required

1. **Solar Panels**: The primary source of power generation in a solar system.

2. **Inverter**: Converts the DC power generated by the solar panels into AC power, which is usable in home appliances.

3. **Mounting Equipment**: Structures that hold solar panels in the optimal position to capture sunlight.

4. **Batteries (for off-grid systems)**: Store excess power generated during the day for use at night or during low sunlight periods.

5. **Charge Controller**: Regulates the voltage and current from the solar panels to the batteries, preventing overcharging and damage.

Selecting Solar Panels

1. **Monocrystalline Panels**:

    o **Efficiency**: Highest efficiency rates (15-20%) because they are made from a single, continuous crystal structure.

    o **Cost**: More expensive due to higher manufacturing complexity.

    o **Application**: Best for areas with limited space due to their higher power output per square foot.

2. **Polycrystalline Panels**:

    o **Efficiency**: Slightly lower efficiency (13-16%) because they are made from fragments of silicon.

    o **Cost**: Less expensive than monocrystalline panels.

     o   **Application**: Suitable for installations with more space, offering a better cost per watt ratio.

Making the Right Choices

- **Space and Budget Considerations**: The choice between monocrystalline and polycrystalline panels often comes down to available space and budget. If space is limited and budget allows, monocrystalline panels are ideal. For larger spaces or tighter budgets, polycrystalline panels are effective.

- **System Scalability**: Consider future needs and whether you may want to expand your system. Ensure components like the charge controller and inverter can handle potential expansion.

Obtaining Permits and Understanding Regulations

Installing a DIY solar system involves navigating various regulations and obtaining necessary permits. This step is essential to ensure that your installation is safe, legal, and eligible for any available incentives. Here's how to proceed with understanding and fulfilling these requirements:

1. Local Regulations

- **Research**: Start by researching local zoning laws, building codes, and solar installation regulations. This information can often be found on local government websites or by contacting your city or county building department.

- **Utility Company Consultation**: If you are planning a grid-tied system, consult your utility company for specific requirements and agreements. They can provide guidance on the technical specifications, grid interconnection processes, and any metering policies for solar systems.

2. Building Permits

- **Application Process**: Contact your local building department to find out the specific documents and plans required for a solar system installation permit. This often includes detailed system designs, electrical schematics, and site plans.

- **Inspection Requirements**: Be prepared for one or more inspections by local authorities. These inspections will check the electrical wiring, mounting of panels, and overall safety of the installation.

- **Fees**: Determine the cost of permit fees in your area as these can vary widely. Including these in your budgeting is essential for the overall project planning.

3. Safety Standards and Compliance

- **National Electrical Code (NEC)**: Ensure your installation complies with the NEC, which sets the benchmark for electrical safety in residential and commercial installations in the U.S.

- **Fire Safety Regulations**: Pay attention to fire safety regulations, particularly regarding the placement of solar panels and the integration with your home's electrical system.

4. Incentives and Rebates

- **Federal and State Incentives**: Investigate available federal and state tax credits, rebates, and other incentives. These can significantly reduce the cost of your solar installation.

- **Local Incentives**: Some local governments offer additional incentives for renewable energy projects. These can include reduced permit fees, expedited permitting processes, or direct rebates.

5. Professional Advice

- **Consulting a Professional**: Although it's a DIY project, consulting with a professional solar installer or an electrician can provide valuable insights, especially in understanding complex regulations and ensuring compliance with all legal and safety standards.

Installation Process for a DIY Solar System

Installing a DIY solar system involves several critical steps, from securely mounting the panels to wiring and grounding the system properly. Here's a detailed guide to help you through each phase of the installation process:

1. Mounting the Solar Panels

- **Choosing a Location**: Ideally, solar panels are mounted on the roof for maximum sun exposure. Ensure that the chosen spot is structurally sound and has minimal shading throughout the day.

- **Installing Mounting Brackets**: Begin by securing the mounting brackets to the roof. The placement should align with your home's rafters for a robust mount. Use sealant to prevent any potential leaks.

- **Setting the Angle**: Adjust the angle of the mounting brackets to optimize the panels for sunlight exposure. This angle depends on your latitude and the time of year.

2. Installing Solar Panels

- **Attaching Panels**: Place the solar panels onto the mounting brackets. Use the hardware provided to securely attach each panel to the mounting system.

- **Alignment**: Ensure that all panels are aligned uniformly. This not only affects the aesthetic but also the performance of your solar array.

3. Wiring the System

- **Connecting Panels**: Connect the solar panels in series or parallel, depending on your voltage and amperage requirements. Use weatherproof wiring and connectors.

- **Running Cables to Inverter**: Route the cables from the solar panels to the inverter's location, typically near your main electrical panel. Ensure that cables are securely fastened and protected from environmental exposure.

- **Battery and Charge Controller**: For off-grid systems, connect the cables from the panels to the charge controller, then from the charge higher controller to the batteries, and finally from the batteries to the inverter.

4. Grounding the System

- **Safety First**: Grounding is crucial for safety and the protection of your system against electrical surges.

- **Grounding Equipment**: Install a grounding rod near the solar array and connect it to the grounding conductor in the solar setup. Ensure all metal components of the solar panels and mounting system are properly grounded.

5. Final Assembly

- **Securing Components**: Double-check all connections, ensuring everything is tightly secured and correctly installed.

- **Protecting the Wiring**: Use conduit or other protective sheathing to protect the wiring from UV, weather, and physical damage.

6. Testing the Installation

- **System Check**: Before fully activating the system, conduct a thorough check to ensure all connections are secure and the system is functioning correctly.

- **Monitoring Setup**: Implement a monitoring system to check the performance of your solar panels and the health of your system regularly.

System Testing and Commissioning

After installing a DIY solar system, the final steps involve thoroughly testing and commissioning the system to ensure it operates safely and efficiently. Here's a detailed guide to help you through these crucial final stages:

1. Checking Connections

- **Visual Inspection**: Begin by visually inspecting all electrical connections. Make sure that all wiring is properly insulated, connections are tight, and there is no visible damage to cables or components.

- **Continuity Testing**: Use a multimeter to check for continuity in the electrical circuits. This test helps verify that there are no open circuits or shorts in the wiring.

- **Safety Standards**: Ensure all installations comply with local electrical codes and safety standards. Check grounding systems and safety disconnects to ensure they are functioning properly.

2. Testing the System

- **Initial Activation**: Gradually power up the system by first connecting the solar panels to the charge controller and then connecting the battery bank (if applicable). Finally, connect the inverter to start converting DC to AC power.

- **Monitoring Output**: Monitor the output of the solar panels using a solar meter to ensure they are producing the expected amount of power. Check the inverter's display panel to verify that it is correctly converting DC power to usable AC power without any fluctuation.

- **Load Testing**: Connect appliances or a load box to test the system's capacity and performance under operational conditions. This helps identify any issues under actual load scenarios.

3. Final Inspection

- **Professional Review**: Depending on local regulations, a final inspection by a certified electric independent might be required to validate the installation. This inspection ensures that the installation meets all local building and electrical codes.

- **Utility Approval**: If the system is grid-tied, the local utility company may need to conduct its inspection and install a net metering device. This process is crucial for integrating your system with the utility grid and enabling electricity billing credits for any surplus energy generated.

- **Documentation**: Keep a record of all inspections, tests, and approvals. This documentation will be crucial for warranties, insurance, and potential resale of the property.

4. Commissioning

- **System Walkthrough**: Once all tests are passed, perform a system walkthrough. Check all components again to ensure they are working as expected.

- **Adjustments**: Make any necessary adjustments to the system settings, such as the charge controller parameters or inverter settings, to optimize performance.

- **Training and Manuals**: Familiarize yourself with the system's maintenance requirements and keep all manuals and manufacturer's guidelines for future reference.

Maintenance of a DIY Solar System

Maintaining a DIY solar system is crucial to ensure its efficiency and longevity. Regular maintenance can prevent potential system failures and maximize the return on your investment. Here are essential maintenance practices for keeping your solar system operating at optimal performance:

1. Regular Monitoring

- **Performance Tracking**: Monitor your system's performance regularly to ensure it is operating within expected parameters. Many modern inverters offer built-in monitoring software that can be accessed via smartphone apps or web platforms. These tools provide real-time data on energy production, system health, and efficiency.

- **Alerts and Notifications**: Set up alerts for any performance anomalies or operational issues. Early detection of drops in power output can indicate potential problems such as panel obstruction or equipment malfunction.

2. Cleaning Panels

- **Scheduled Cleaning**: Develop a routine cleaning schedule for your solar panels. In most regions, cleaning them twice a year is sufficient, but in areas with high dust or pollen, more frequent cleaning may be necessary.

- **Proper Methods**: Use a soft brush or a cloth with mild soapy water to clean the panels. Avoid harsh chemicals or abrasive materials that could damage the panel surface. For panels installed in hard-to-reach areas, consider hiring professionals or using automated cleaning systems.

- **Check for Obstructions**: Regularly inspect the panels for debris, snow, or bird droppings, which can block sunlight and reduce panel efficiency.

3. Routine Checks

- **Visual Inspections**: Perform a visual inspection of all system components periodically. Look for signs of wear and tear, such as cable insulation deterioration, rust on mounting equipment, or loose fittings and fasteners.

- **Connection Integrity**: Check all electrical connections for tightness and integrity. Loose connections can lead to arcing and potentially cause fires.

- **Protective Measures**: Inspect protective devices such as surge protectors and circuit breakers to ensure they are functional. Replace any that show signs of damage or wear.

4. Professional Assessments

- **Annual Inspections**: Although many aspects of solar system maintenance can be DIY, it's advisable to have the system checked annually by a professional. They can perform detailed checks and tests that are more technical in nature, such as inverter efficiency tests and thermal imaging to detect hot spots in electrical connections.

- **Documentation and Records**: Keep detailed records of all maintenance activities, repairs, and professional inspections. This documentation can be invaluable for warranty claims, selling your home, or simply keeping track of the system's history.

Benefits of a DIY Solar System

Installing a DIY solar system offers several distinct advantages, from financial benefits to personal satisfaction and environmental impact. Here's an in-depth look at the key benefits:

1. Cost Savings

- **Reduced Upfront Costs**: One of the most significant benefits of a DIY solar system is the potential cost savings. By managing the installation yourself, you can save on the labor costs typically associated with professional installations. This can reduce the overall investment required to go solar by a substantial margin.

- **Control Over Expenses**: DIY projects allow you to shop around for the best deals on materials and components. You can choose when and where to buy, potentially taking advantage of sales and discounts that professional installers may not utilize.

2. Learning Experience

- **Understanding Solar Technology**: Installing your own solar system provides a hands-on opportunity to learn about solar technology. You'll gain insight into how photovoltaic cells generate electricity, how different components like inverters and charge controllers work, and how to optimize the system for maximum efficiency.

- **Skills Development**: Beyond just learning about solar technology, you'll develop practical skills in electrical wiring, system design, and general construction techniques. These skills are not only useful for maintaining your solar system but can also be applied to other home improvement projects.

3. Energy Independence

- **Self-Sufficiency**: A DIY solar system can increase your household's self-sufficiency by reducing or even eliminating your reliance on the grid. This is particularly advantageous during power outages or in areas with unreliable electricity supply.

- **Protection Against Energy Price Fluctuations**: By generating your own electricity, you protect yourself from the volatility in energy prices. This can lead to predictable electricity costs and, over time, significant savings on your energy bills.

4. Environmental Impact

- **Reducing Carbon Footprint**: Solar energy is a clean, renewable resource that reduces your household's carbon footprint. By installing a solar system, you are directly contributing to the reduction of greenhouse gas emissions.

- **Promoting Renewable Energy**: DIY solar installations help promote the adoption of renewable technologies by demonstrating their viability and encouraging others in your community to consider solar.

5. Enhanced Property Value

- **Increased Home Value**: Homes with solar systems often have a higher market value and can be more attractive to buyers looking for energy-efficient properties.

- **Attract Eco-Conscious Buyers**: A DIY solar system can be a selling point for potential homebuyers who are particularly interested in sustainability and self-sufficiency.

Wind Energy

Wind energy is a renewable power source that utilizes the force of the wind to generate electricity. This section explores the fundamentals of wind energy, including how it works, its applications, and the contexts in which it can be most effectively used.

Description of Wind Energy

Wind energy stands as a pillar of renewable energy solutions by harnessing the natural power of wind to generate electricity. This form of energy conversion is not only sustainable but also increasingly important in the global shift toward green energy sources. Below is a detailed exploration of how wind energy is harnessed, the components involved, and the factors that influence its efficiency.

Kinetic Energy Conversion

- **Process**: Wind turbines convert the kinetic energy present in wind into mechanical power. This transformation occurs when wind passes over the turbine's blades, causing them to rotate. The rotation of the blades turns a connected shaft inside the nacelle, which is the turbine's main body sitting atop the tower.

- **Electricity Generation**: The shaft turns a generator, which then converts this mechanical power into electricity. The electricity produced is initially in the form of alternating current (AC) or direct current (DC), depending on the design of the generator.

Components of a Wind Turbine

- **Rotor (Blades and Hub)**: The rotor comprises the blades and the hub to which they are attached. The blades are designed to capture wind energy efficiently, and their number can vary, although most modern turbines have three blades.

- **Drive System**: The drive system typically includes a gearbox and a generator. The gearbox adjusts the rotational speed from the rotor to suit the generator's required speed to produce electricity efficiently.

- **Generator**: Converts the mechanical energy from the rotor into electrical energy. Some turbines use direct drive systems that eliminate the need for a gearbox, reducing maintenance and mechanical complexity.

- **Tower**: Supports the nacelle and rotor and elevates them to heights where winds are stronger and less turbulent.

- **Nacelle**: Houses the generator, gearbox, drive shaft, and other components. It is positioned on top of the tower and includes mechanisms to control the turbine's operation.

Efficiency Factors

- **Wind Speed**: The amount of energy a wind turbine can generate largely depends on the wind speed; power output increases exponentially with wind speed. Turbines typically start generating electricity at wind speeds of around 4-5 meters per second and reach maximum capacity at about 12-15 meters per second.

- **Air Density**: Affects the amount of kinetic energy available in the wind. Colder, denser air contains more energy than warmer, less dense air at the same speed.

- **Turbine Design**: The length and pitch of the blades significantly influence efficiency. Longer blades can capture more wind energy, and adjustable pitch allows the blades to rotate to capture the optimal amount of energy across varying wind speeds.

- **Site Selection**: Choosing the right location is crucial for maximizing efficiency. Ideal sites have strong, consistent winds, and minimal obstructions such as buildings or trees that cause turbulence.

Site Considerations

- **Environmental and Visual Impact**: Wind farms are sometimes opposed due to their visual impact and potential effects on local wildlife, especially birds and bats. Careful site selection and environmental assessments can mitigate these impacts.

- **Community Engagement**: Engaging with local communities and stakeholders early in the planning process can address concerns and enhance support for new projects.

Applications of Wind Energy

Wind energy is a versatile and increasingly vital source of renewable energy that serves various purposes, from bolstering national grids to providing independent power in remote locations. Below, we explore the different applications of wind energy that demonstrate its adaptability and crucial role in a sustainable energy future.

Large-Scale Power Generation

- **Wind Farms**: Large wind farms are the most common manifestation of wind energy application. These installations consist of dozens to hundreds of turbines lined up in areas known for strong and consistent winds. They can be located onshore (land-based) or offshore (in bodies of water).

- **Grid Connection**: Electricity generated from these wind farms is typically fed into the national grid, where it mixes with power from other sources. By supplying electricity on such a scale, wind farms can significantly reduce a region's reliance on non-renewable energy sources, leading to reductions in greenhouse gas emissions.

- **Economic Impact**: Besides generating electricity, large-scale wind projects can stimulate local economies through job creation in construction, maintenance, and ongoing operations. They often also generate income for landowners and local governments through taxes and lease payments.

Distributed Power for Remote Locations

- **Off-Grid Solutions**: Wind turbines are an effective solution for generating electricity in remote areas where extending the central power grid would be prohibitively expensive or logistically challenging. These setups can be tailored to meet the specific needs of a community or facility, providing a steady and reliable power supply.

- **Hybrid Systems**: In some cases, wind power is combined with other renewable sources, such as solar energy and diesel generators, to create hybrid systems. These systems can provide more consistent energy output, adapting to changes in wind speed and solar availability.

Supplementing Existing Grids

- **Grid Stability and Diversification**: Integrating wind energy into existing power grids helps diversify energy sources, which can enhance grid stability. Wind energy can provide additional power during periods of high demand or when other power plants are offline.

- **Reducing Carbon Footprints**: By supplementing grids with wind energy, regions can lower their carbon emissions, contributing to national and international goals for reducing the impact of climate change.

- **Peak Shaving**: Wind energy can be particularly useful for meeting peak energy demands, especially in combination with energy storage systems that store excess wind energy and discharge it when demand peaks.

Innovations and Expansions

- **Technological Advancements**: The ongoing improvements in turbine technology, such as enhanced blade designs and more efficient generators, continue to increase the viability and efficiency of wind energy.

- **Future Prospects**: As technology progresses and investment continues, wind energy is expected to play an even more integral role in global energy strategies, especially as countries aim to meet ambitious renewable energy targets.

Installation Locations for Wind Energy

Wind energy projects can be implemented in a variety of locations, each with unique characteristics that influence the efficiency, cost, and potential environmental impact of the installation. Here's an overview of the primary locations for installing wind farms:

Onshore Wind Farms

- **General Characteristics**: Onshore wind farms are situated on land, often in open plains, hilltops, or mountain passes where wind speeds are higher. These locations are selected based on their accessibility to wind resources and proximity to existing power grids to facilitate easier energy distribution.

- **Advantages**:

  o **Cost-Effectiveness**: The cost of constructing and maintaining onshore wind farms is generally lower than that of offshore installations. This is due to easier accessibility, lower logistics and foundation costs, and simpler maintenance procedures.

  o **Ease of Installation**: Onshore sites are more accessible for transporting large turbine components, and the infrastructure such as roads and electricity networks is usually better developed than in offshore locations.

  o **Quicker Construction**: Projects on land typically face fewer regulatory hurdles and can be completed faster, allowing them to begin generating electricity sooner.

- **Challenges**:

  o **Visual and Noise Impact**: Onshore wind farms can be controversial due to their visual impact on landscapes and potential noise, which can lead to opposition from local communities.

  o **Variable Wind Speeds**: While many onshore locations experience strong winds, the variability and obstruction caused by terrain features can affect performance.

Offshore Wind Farms

- **General Characteristics**: These installations are built in bodies of water, such as lakes, seas, or oceans. Offshore wind farms take advantage of the open environment, which typically offers more consistent and stronger winds than on land.

- **Advantages**:

  o **Higher Energy Output**: Offshore winds tend to be faster and more stable, which can significantly increase a turbine's energy output compared to onshore settings.

- o **Reduced Impact on Communities**: Situated away from populated areas, offshore wind farms minimize visual and noise impacts, making them more acceptable to the public.

- **Challenges**:
  - o **Higher Costs**: The costs associated with constructing and maintaining offshore wind farms are considerably higher due to the need for specialized ships, more robust turbine designs, and complex foundations that can withstand marine conditions.

  - o **Logistical and Maintenance Challenges**: Accessing turbines for maintenance is more difficult and weather-dependent, which can increase the time and cost of operations.

  - o **Environmental and Navigational Considerations**: Installing turbines in marine environments requires careful assessment of impacts on marine life and navigation, which involves extensive environmental reviews and compliance with maritime regulations.

Future Prospects

- **Technological Innovations**: Advances in technology are gradually overcoming the challenges associated with offshore wind installations. Floating turbines, for instance, are opening up new areas for wind energy generation where deep waters made traditional fixed-foundation turbines unfeasible.

- **Expansion Potential**: As the demand for renewable energy grows, both onshore and offshore wind farms are expected to expand, with offshore projects gaining traction due to their enormous potential for energy production.

Benefits of Wind Energy

Wind energy is increasingly recognized as a pivotal solution for a sustainable future, offering numerous benefits ranging from environmental to economic. Here are the key advantages of integrating wind energy into our energy mix:

Environmental Benefits

- **Zero Emissions**: Wind turbines produce electricity without emitting greenhouse gases, which is crucial in mitigating climate change. Unlike fossil fuels, wind energy does not release pollutants like carbon dioxide or sulfur dioxide, making it a clean and environmentally friendly energy source.

- **Reduced Dependence on Fossil Fuels**: By increasing the share of wind energy in the power generation mix, countries can reduce their dependence on coal, oil, and natural gas. This shift not only conserves finite natural resources but also reduces the environmental degradation associated with mining and drilling.

- **Minimal Water Use**: Wind energy generation does not require water, which is a significant benefit in areas suffering from water scarcity. In contrast, traditional power plants often require large amounts of water for cooling processes, placing a strain on local water resources.

Sustainability

- **Renewable Resource**: Wind is a plentiful and inexhaustible natural resource. As long as the wind continues to blow, it can be harnessed to generate electricity, providing a perpetual energy supply.

- **Scalability and Versatility**: Wind turbines can be installed in a variety of settings, including rural, urban, onshore, and offshore locations. This versatility allows for scalable solutions that can be tailored to meet regional energy demands without the environmental footprint associated with larger infrastructure projects.

Cost-Effectiveness

- **Low Operational Costs**: Once a wind turbine is installed, the costs of operation and maintenance are relatively low compared to those of conventional power plants. Wind energy does not require any fuel costs, which significantly reduces its lifetime operational expenses.

- **Economies of Scale**: As the wind energy sector grows and technology advances, the costs associated with manufacturing, installing, and maintaining wind turbines continue to decrease. This trend, known as economies of scale, has made wind energy more competitive with traditional energy sources.

- **Job Creation**: The wind energy industry is labor-intensive, particularly in the manufacturing, installation, and maintenance stages. This creates jobs and stimulates economic activity in local communities, contributing to economic development in areas hosting wind projects.

Energy Security

- **Diversification of Energy Portfolio**: Incorporating wind energy into a region's energy portfolio enhances energy security by diversifying the sources of electricity generation. This reduces vulnerability to energy supply disruptions and price volatility associated with fossil fuels.

- **Local Energy Generation**: Wind energy can be generated locally, reducing the need for extensive transportation and distribution networks. This not only decreases energy loss in transmission but also improves the resilience of the energy system against external disruptions.

Challenges of Wind Energy

While wind energy presents numerous benefits as a sustainable and clean power source, there are several challenges associated with its widespread adoption. These obstacles can impact the efficiency, community acceptance, and overall sustainability of wind energy projects.

Intermittency and Reliability

- **Variable Nature**: Wind energy is inherently intermittent; the wind doesn't blow consistently at all times. This variability can lead to fluctuations in energy production, which can be challenging for maintaining the stability of the power grid.

- **Energy Storage Needs**: To address the intermittency, substantial investments in energy storage technologies are necessary. Systems such as batteries or pumped hydro storage can store excess energy generated during windy periods and release it during calm periods, but these solutions add to the cost and complexity of wind energy projects.

- **Backup Systems**: Grid operators often need to have backup power generation systems, typically from more consistent energy sources like natural gas, to ensure a steady energy supply when wind production is low. This requirement can detract from the environmental benefits of wind energy if fossil fuels are used as the backup.

Community and Environmental Impact

118

- **Noise Concerns**: Wind turbines generate a significant amount of noise, which can be a nuisance for nearby residents. The sound of the rotating blades and mechanical noise from the turbine can lead to community opposition.

- **Visual Impact**: The presence of large wind turbines can significantly alter the visual landscape, which may be viewed negatively by communities, especially in scenic or rural areas. The aesthetic considerations can lead to resistance against the installation of wind farms.

- **Wildlife Interactions**: Wind turbines pose a threat to wildlife, particularly birds and bats, which can collide with turbine blades. The environmental impact on local ecosystems can be significant, leading to further scrutiny and regulatory challenges for new wind projects.

- **Land Use**: While wind turbines themselves occupy relatively small footprints, they need to be spaced widely apart to function efficiently. This can lead to large areas of land being used for wind farms, which might compete with other land uses such as agriculture or conservation efforts.

Technical and Infrastructure Challenges

- **Site Specificity**: The efficiency of wind turbines heavily depends on local wind speeds and uninterrupted airflow, which means that suitable sites are often located in remote areas that may lack existing infrastructure. Developing these sites can be expensive and logistically challenging.

- **Transmission Requirements**: Generating wind power in remote locations necessitates extensive and often expensive transmission lines to connect the electricity to where it is needed in urban centers. Building these transmission networks can be as costly and controversial as the turbines themselves.

Economic and Policy Hurdles

- **Capital Costs**: The initial capital investment for wind turbines and the necessary infrastructure is significant. While operational costs are low, the upfront expenses can be a barrier to entry for some developers.

- **Policy Dependence**: The development and expansion of wind energy often rely heavily on government policies and incentives. Changes in policy or reductions in subsidies can abruptly affect the viability and attractiveness of wind energy projects.

Addressing these challenges requires a multifaceted approach, including technological advancements, strategic planning, community engagement, and supportive policies. By overcoming these hurdles, wind energy can continue to grow as a key component of the global renewable energy portfolio, contributing to a sustainable and secure energy future.

Book 7: Alternative Energy Sources

Implementing Biogas Systems for Energy and Heat Production

Biogas systems, also known as anaerobic digestion systems, offer a sustainable solution for producing energy and heat from organic materials. These systems are increasingly recognized for their ability to convert agricultural waste, food scraps, and other organic materials into a usable form of renewable energy. Here's an in-depth look at how to implement a biogas system for energy and heat production:

Understanding Biogas Biogas is a mixture of methane (CH_4) and carbon dioxide (CO_2), produced through the anaerobic digestion of organic matter. This process occurs in an oxygen-free environment where microorganisms break down organic materials such as animal waste, plant material, and food scraps. The primary output, methane, is a potent energy source, while carbon dioxide is a byproduct. Additionally, the process generates digestate, a nutrient-rich substance that can greatly benefit agricultural applications as a high-quality fertilizer.

Benefits of Biogas Systems

1. **Renewable Energy Production** Biogas systems efficiently convert a diverse range of organic wastes into methane, a clean and renewable energy source. This methane can be utilized to generate electricity and provide heating solutions, significantly reducing reliance on fossil fuels and enhancing energy security.

2. **Waste Reduction** By processing organic wastes, biogas systems reduce the overall volume of waste that must be handled by landfills. This reduction plays a crucial role in decreasing the emission of landfill

gases, potent greenhouse gases that contribute to global warming. Furthermore, diverting organic waste to biogas production helps mitigate environmental pollution and extends the lifespan of landfills.

3. **Soil Enhancement** The digestate produced from biogas plants is a nutrient-rich byproduct, making it an excellent biofertilizer. It enriches the soil with essential nutrients, which can improve soil fertility and structure. Using digestate reduces the dependence on chemical fertilizers, promoting a more sustainable form of agriculture that is less harmful to the environment.

Comprehensive Impact Biogas systems not only provide a practical solution to waste management challenges but also contribute to sustainable agriculture and energy production. They offer a circular economy approach where waste products are converted into valuable energy and agricultural inputs, thereby closing the loop and reducing environmental impacts. As such, the implementation of biogas technology is increasingly recognized as a key component in achieving renewable energy targets and sustainable waste management strategies.

Steps for Implementing a Biogas System

Feasibility Study for Implementing Biogas Systems:

Conducting a feasibility study is a crucial first step in the implementation of a biogas system. This study evaluates various aspects to ensure the viability and potential success of the project. Here's what it entails:

1. **Assessment of Organic Waste Availability**

   o **Identify Organic Waste Sources**: Begin by cataloging potential sources of organic waste within the area, such as agricultural residues, food waste from local businesses, and waste products from livestock facilities.

   o **Quantify Waste**: Estimate the total available organic waste by gathering data on the waste generation rates from these sources. This will help determine the potential biogas yield and the scale at which the plant should operate.

   o **Analyze Waste Composition**: Different types of organic waste produce varying amounts of biogas. Conduct laboratory tests to analyze the biogas potential of the available organic waste types. This analysis will be critical in designing the system to maximize output and efficiency.

2. **Site Analysis**

   o **Proximity to Waste Sources**: Select a site that is close to major sources of organic waste to minimize the cost and complexity of waste transportation. This proximity is vital for maintaining a sustainable and economically viable operation.

   o **Space Requirements**: Evaluate the physical space available for installing a biogas digester and associated infrastructure, such as storage tanks and energy conversion units. Consider future expansion possibilities in the initial site layout.

   o **Environmental Compliance**: Check local zoning laws and environmental regulations to ensure that the site meets all legal requirements for waste handling and biogas production. This includes obtaining necessary permits and ensuring that the operation will not adversely affect the local ecosystem.

- o **Infrastructure Access**: Assess the availability of necessary infrastructure, such as roads for transport, water supply for the digestion process, and electrical grids for connecting the energy production facilities. The presence of these infrastructures can significantly influence the feasibility and operational costs of the biogas system.

Conducting a thorough feasibility study not only ensures that the biogas project is technically and economically viable but also helps in laying the groundwork for a successful implementation. This initial step is essential for aligning the project with environmental goals and regulatory frameworks, thereby securing the sustainability of the project in the long run.

System Design and Technology Selection for Biogas Systems

Designing a biogas system involves careful consideration of the type of digester, sizing the system appropriately, and integrating it with existing energy systems. Here's a detailed guide to help you through the process:

1. **Type of Digester**

   - o **Covered Lagoon Digesters**: Best suited for milder climates where the temperature does not drop too low, these digesters are economical and simple. They work well with dilute waste streams such as dairy effluent.

   - o **Complete Mix Digesters**: These are enclosed tanks that maintain a controlled environment, ideal for colder climates. They can handle a variety of organic wastes, including food waste and manure, by mixing them consistently to enhance gas production.

   - o **Plug Flow Digesters**: Suitable for more solid wastes with a dry matter content higher than 11%. This type of digester typically operates in linear tanks where waste moves in one direction, making it efficient for farms with substantial livestock waste.

2. **Sizing the System**

   - o **Assess Daily Waste Input**: Estimate the amount of organic waste that can be consistently supplied to the digester daily. This assessment helps in determining the capacity of the digester needed.

   - o **Biogas Output Requirements**: Based on your energy needs, calculate the required daily biogas production. This will guide the overall size of the biogas system, including storage and processing facilities.

   - o **Scale and Expansion**: Design the system with scalability in mind to accommodate future increases in organic waste availability or energy demands.

3. **Integration with Existing Energy Systems**

   - o **Direct Combustion**: If the biogas is intended primarily for heating, design the system to include a combustion system that can utilize biogas efficiently.

   - o **Electricity Generation**: For systems aimed at electricity production, include a biogas generator in your design. Determine the generator capacity based on the expected biogas output and the electricity demand.

- **Upgrading to Biomethane**: If the biogas is to be upgraded to biomethane for injection into gas grids, include the necessary technology for gas cleaning and upgrading in your design. Consider the specifications of the local gas grid to ensure compatibility.

- **Heat Recovery Systems**: Implement heat recovery from the biogas combustion process to enhance overall system efficiency. This can be used to heat buildings or the digester itself during colder months.

By meticulously planning the type of digester, sizing the system, and determining how the produced biogas will be used, you can maximize the efficiency and sustainability of the biogas system. This strategic approach not only ensures a reliable energy source but also promotes the integration of renewable energy into existing infrastructures.

Construction and Installation of Biogas Systems

Constructing and installing a biogas system involves building the digester and setting up the necessary equipment for gas collection, processing, and utilization. Here's how to effectively approach the construction and installation phase:

1. **Building the Digester**

   - **Material Selection**: Choose materials that are durable and resistant to corrosion, such as stainless steel or specially coated metals, to ensure the longevity of the digester. The construction material should also be impervious to the acidic conditions created by the fermentation process.

   - **Structural Integrity**: Design the digester to withstand both internal and external pressures and potential aggressive environments. This may include reinforcing the structure to prevent collapse and leakage.

   - **Insulation**: Properly insulate the digester to maintain the necessary temperature for optimal microbial activity, especially in climates with significant temperature variations.

2. **Installation of Gas Collection and Utilization Equipment**

   - **Gas Collection System**: Install a reliable system to capture biogas produced in the digester. This includes gas domes or covers that prevent leaks and capture gas efficiently.

   - **Gas Cleaning and Compression**: Set up equipment to remove impurities such as hydrogen sulfide, moisture, and particulates from the biogas. This step is crucial for protecting downstream equipment and ensuring efficient operation. Compression systems are also necessary to increase the pressure of biogas for storage or pipeline injection.

   - **Gas Storage**: Install adequate storage facilities to hold biogas until it is needed, allowing for flexible and efficient energy production. Gas storage systems should be designed to handle the pressure and composition of biogas safely.

   - **Energy Conversion Equipment**:
     - **Boilers**: For systems focusing on heating, install biogas-compatible boilers that can convert biogas into thermal energy efficiently.

- **Turbines and CHP Systems**: For electricity generation, turbines or combined heat and power systems can be used. CHP systems are particularly efficient as they can produce both heat and electricity from the biogas, significantly increasing the overall energy efficiency of the system.

3. **System Integration**

   o **Integration with Existing Infrastructure**: Ensure the biogas system is seamlessly integrated with existing energy systems, allowing for smooth operation and minimal disruption.

   o **Control Systems**: Implement automated control systems to monitor and adjust the flow of biogas and the operation of energy conversion equipment. This ensures optimal performance and safety of the system.

By carefully constructing the digester and installing high-quality equipment for gas collection and utilization, a biogas system can provide a reliable and sustainable source of energy. Proper planning and execution during this phase are critical to the success and efficiency of the entire system.

Operation and Maintenance of Biogas Systems

The successful operation and ongoing maintenance of biogas systems are crucial for ensuring their longevity and efficiency. Here's a detailed guide to help manage the daily operations and upkeep of a biogas system.

Start-up Process

- **Establishing Microbial Community**: The initial phase involves creating a conducive environment for the microbes that will break down the organic material. This usually starts with inoculating the digester with a seed slurry from another active digester or a special microbial culture designed for biogas production.

- **Gradual Loading**: Begin by gradually adding organic material to avoid overwhelming the system. This allows the microbial community to adapt and grow, stabilizing the digestive process.

- **Monitoring Adjustments**: Keep a close eye on key parameters such as temperature, pH, and gas production during this phase. Adjustments may be necessary to maintain an optimal environment for the microbes.

Regular Feeding and Monitoring

- **Consistent Feeding Schedule**: Feed the digester with a consistent amount of organic waste daily to maintain continuous gas production. The type and amount of feedstock directly influence the quantity and quality of biogas produced.

- **pH and Temperature Checks**: Regularly check the pH and temperature of the digester content. The ideal pH range for anaerobic digestion is typically between 6.8 and 7.2, and the temperature will depend on whether the system is mesophilic (around 37°C) or thermophilic (around 55°C).

- **Biogas Monitoring**: Monitor the biogas production rate and composition (methane and carbon dioxide levels) to assess the health and efficiency of the digestion process. Sudden changes in gas production can indicate issues such as overloading or toxic substance contamination.

Maintenance

- **Routine Inspections**: Regularly inspect physical components of the biogas system, including pipes, seals, and valves, for signs of wear and tear or leaks. Preventive maintenance can help avoid unexpected breakdowns and gas leaks.

- **Cleaning and Repairs**: Periodically clean and remove built-up residue inside the digester and pipelines to prevent blockages. Repair or replace damaged parts to ensure the system functions efficiently.

- **Safety Checks**: Perform safety inspections to check the structural integrity of gas storage tanks and ensure that all safety valves and fire protection systems are functioning correctly.

Long-term Health of the System

- **Sludge Removal**: Over time, digestate accumulates in the digester. Regularly remove excess sludge to prevent reduction in the digester's capacity and efficiency.

- **Documentation and Record Keeping**: Keep detailed records of operation parameters, maintenance activities, and any issues or repairs. This documentation can be invaluable for troubleshooting problems and planning future upgrades or expansions.

- **Training and Capacity Building**: Ensure that staff responsible for operating and maintaining the biogas plant are properly trained and knowledgeable about the latest practices and safety standards.

Effective operation and maintenance are key to maximizing the output and lifespan of a biogas system. By adhering to these guidelines, operators can ensure that their biogas plants remain reliable, efficient, and safe over the years.

Regulatory Compliance and Safety Measures for Biogas Systems

Proper adherence to regulatory compliance and safety protocols is essential when managing a biogas system. This not only ensures legal operation but also protects the equipment, environment, and personnel involved. Here's an exhaustive guide on navigating these critical aspects.

Permits and Approvals

- **Local and Environmental Permits**: Before construction and operation commence, obtain all necessary permits from local government bodies. This may include environmental permits, particularly if the biogas plant has potential impacts on local ecosystems or water sources.

- **Building and Zoning Approvals**: Ensure that the biogas system complies with local building codes and zoning requirements. This involves securing construction permits and possibly undergoing inspections to certify that the installation meets all structural and safety standards.

- **Health and Safety Regulations**: Check for any health and safety regulations that apply specifically to biogas production, especially concerning the handling and processing of organic waste.

Safety Protocols

- **Gas Leak Prevention and Detection**: Install gas detectors to monitor for methane leaks, which can pose serious safety risks. Regularly check and maintain all gas pipes and connections to prevent leaks.

- **Ventilation Systems**: Implement adequate ventilation systems in areas where biogas is processed or stored to prevent gas accumulation and reduce the risk of explosions.

- **Emergency Response Plan**: Develop and maintain an emergency plan that includes procedures for dealing with gas leaks, fires, and other potential accidents. Regularly conduct drills to ensure that all workers are familiar with the plan and can respond effectively in an emergency.

- **Protective Gear**: Provide appropriate personal protective equipment (PPE) to all personnel involved in the operation of the biogas plant. This should include gloves, eye protection, and masks or respirators suitable for working in environments where gas may be present.

- **Chemical Handling Safety**: Since the digestate and other by-products may contain harmful bacteria or chemicals, ensure safe handling and storage practices are in place. This includes proper training for workers on how to handle these materials safely and the use of suitable containers for storage and transport.

Compliance Monitoring

- **Regular Audits and Inspections**: Conduct regular audits and inspections to ensure ongoing compliance with all regulatory and safety requirements. This may involve both internal audits and inspections by external authorities.

- **Documentation and Record Keeping**: Keep comprehensive records of compliance efforts, safety inspections, and any incidents or breaches. These records are crucial for demonstrating adherence to regulations during inspections and for improving safety protocols.

By rigorously applying these regulatory and safety measures, biogas facilities can maintain a safe working environment while ensuring compliance with local and national standards. This proactive approach minimizes risks and enhances the overall efficiency and sustainability of biogas production.

Conclusion

Implementing a biogas system provides a practical and environmentally friendly way to manage organic waste while producing valuable energy and heat. With careful planning, technical knowledge, and regulatory compliance, biogas systems can contribute significantly to sustainable energy goals, offering a reliable source of renewable energy for various applications.

Practical Use of Hand-Crank Generators and Other Low-Tech Solutions

In today's high-tech world, there remains a significant role for low-tech solutions in sustainable living and emergency preparedness. Among these, hand-crank generators stand out for their simplicity and reliability. Here's an exhaustive guide on how to effectively utilize these devices and other similar low-tech solutions.

Description and Operation

- **Mechanism**: Hand-crank generators function by turning a handle manually, which rotates a dynamo or alternator. This mechanical motion converts kinetic energy into electrical energy through electromagnetic induction.

- **Energy Output**: The electricity generated is typically direct current (DC) and is used immediately to power connected devices or to charge batteries. The amount of power generated depends on the speed and consistency of the crank operation.

- **Components**: The primary components include a handle, a gear system to increase revolutions, a dynamo or alternator for electricity production, and often a built-in battery for storing electricity.

Applications

- **Emergency Kits**: In emergencies like natural disasters when the power grid may be down, hand-crank generators provide an essential source of power for charging mobile phones, radios, and other small communication devices.

- **Camping Trips**: For outdoor enthusiasts, these generators are a reliable power source for charging GPS devices, portable lights, and small cooking gadgets, enhancing off-grid camping experiences.

- **Remote Areas**: In remote locations where traditional power infrastructure is either unavailable or unreliable, hand-crank generators serve as a critical tool for basic electricity needs.

- **Disaster Recovery**: During recovery efforts after disasters, these generators can power essential communication equipment and lighting, supporting rescue and relief operations.

Benefits of Hand-Crank Generators

Hand-crank generators provide a reliable, portable, and environmentally friendly alternative for generating electricity, particularly useful in emergencies, remote locations, or off-grid activities. Here are the primary benefits outlined in detail:

Reliability

- **Immediate Power Source**: Hand-crank generators are an extremely reliable source of power. They do not depend on external conditions such as sunlight, making them highly effective in all weather conditions, day or night.

- **Always Ready**: These generators are always ready for use with minimal preparation. There's no need for fuel or sunlight; simply turn the crank to generate electricity.

Portability

- **Compact Design**: Hand-crank generators are designed to be compact and lightweight. This design makes it easy to pack them in emergency kits, carry on camping trips, or store in small spaces at home or in a vehicle.

- **Ease of Transportation**: Their size and weight make hand-crank generators particularly useful for situations where carrying large amounts of equipment is impractical, such as hiking or emergency evacuations.

Eco-Friendly

- **Zero Emissions**: One of the most significant benefits of hand-crank generators is that they operate without producing any pollutants. Since they require only human power, they do not emit greenhouse gases or pollutants that are typical of fossil fuel-powered generators.

- **Sustainable Operation**: These generators contribute to sustainable living practices by offering an alternative to conventional power sources that deplete natural resources and harm the environment.

Additional Benefits

- **Health and Fitness**: Using a hand-crank generator can also serve as a physical activity, potentially providing a minor workout akin to using a stationary bicycle.

- **Independence from Grid Power**: For those seeking independence from traditional power sources, hand-crank generators provide a degree of self-sufficiency, reducing reliance on public utilities.

- **Cost-Effective**: After the initial purchase, there are no ongoing fuel costs, which can provide significant savings over time compared to generators that require gasoline or diesel.

Overall, hand-crank generators are a practical solution for anyone needing a dependable, mobile, and green source of power. They are particularly valuable for emergency preparedness, providing peace of mind that power will be available when it's most needed, regardless of external conditions.

Other Low-Tech Energy Solutions

In addition to hand-crank generators, there are several other low-tech solutions that can provide sustainable and efficient energy sources. These alternatives harness natural forces and human effort to create usable energy without relying on advanced technology or non-renewable resources. Here's a closer look at some of these innovative solutions:

Solar Cookers

- **Operation**: Solar cookers use the sun's rays, captured and concentrated through reflective materials, to cook food or boil water. They typically consist of an insulated box with a transparent lid and reflective panels to direct sunlight into the box.

- **Benefits**: Solar cookers are eco-friendly, utilizing a renewable resource without emitting pollutants. They are especially useful in sunny climates and can significantly reduce the need for traditional cooking fuels like wood, charcoal, or gas, helping to conserve these resources and reduce air pollution.

- **Applications**: Ideal for outdoor activities like camping and suitable for use in remote areas where traditional energy sources are scarce or expensive. They can also be used during power outages or in disaster response scenarios to provide a means of cooking without electricity.

Water Wheel Chargers

- **Operation**: Water wheel chargers generate electricity through a simple mechanical setup where a wheel is turned by flowing water, much like traditional water mills. This motion drives a turbine or a connected generator to produce electricity.

- **Benefits**: This system is sustainable and efficient in areas with suitable water flows. It harnesses a free natural resource without affecting the water's ecosystem and operates continuously as long as there is flowing water.

- **Applications**: Particularly useful in rural or undeveloped areas with access to flowing streams or rivers. These systems can power small appliances, charge batteries, and provide lighting, contributing significantly to the energy needs of off-grid communities.

Bicycle Generators

- **Operation**: Bicycle generators convert the kinetic energy from pedaling into electrical energy. This is typically achieved by attaching a dynamo to a stationary bicycle, which produces electricity as the user pedals.

- **Benefits**: Bicycle generators provide a dual benefit of generating electricity while promoting physical fitness. They are an excellent emergency power source, requiring only human energy to function.

- **Applications**: These generators are ideal for maintaining a degree of electricity supply during power outages, in off-grid situations, or simply for exercise purposes. They can charge batteries, power lights, and run small electronic devices.

Implementing Low-Tech Solutions

- **Community and Educational Programs**: Introducing these technologies in community centers and schools can help promote sustainable practices and educate the public on alternative energy sources.

- **Integration with High-Tech Systems**: While low-tech, these solutions can be integrated with more complex systems to provide a reliable and diverse energy mix, enhancing overall resilience to energy supply issues.

- **Customization and Innovation**: These systems can often be built or modified with locally available materials, allowing for innovation and customization to meet specific local needs.

These low-tech energy solutions offer significant benefits, particularly in terms of sustainability, independence from traditional energy grids, and potential cost savings. They exemplify how simple, time-tested technologies can still play a crucial role in our modern energy landscape.

Maintenance and Care of Low-Tech Energy Devices

Proper maintenance and care are crucial for the longevity and efficiency of low-tech energy solutions like hand-crank generators, solar cookers, water wheel chargers, and bicycle generators. Here's how to ensure these devices continue to operate effectively:

Regular Checks

- **Lubrication**: Regularly lubricate all moving parts to reduce friction and wear. This is particularly important for devices with mechanical components such as bicycle generators and water wheel chargers.

- **Corrosion Checks**: Inspect all metal parts for signs of corrosion, especially for devices exposed to outdoor elements like solar cookers and water wheels. Applying a rust-resistant coating can help mitigate this issue.

- **Component Replacement**: Replace worn-out or damaged parts promptly. Check connections and cables for signs of wear or damage, and ensure that all electrical components are in good condition.

Storage

- **Environment**: Store all devices in a dry, cool place when not in use. Avoid places with high humidity or extreme temperatures, which can damage the components.

- **Protective Covering**: Use covers for larger installations like solar cookers and water wheels to protect them from dust, debris, and weather when not in active use.

- **Organized Space**: Keep all related accessories, such as cables, tools, and spare parts, stored together with the device for easy access during maintenance or emergency use.

Challenges and Considerations

While low-tech energy solutions offer many benefits, they also come with challenges that must be considered before implementation:

Physical Effort

- **Sustainability of Effort**: Devices like hand-crank generators require continuous physical effort to generate power, which may not be practical for everyone, especially for those with physical limitations or during emergencies.

- **Fatigue and Efficiency**: Extended use of manual energy devices can lead to fatigue, reducing the efficiency of energy production. It's important to assess the practicality based on the physical capabilities of the users.

Power Limitations

- **Energy Output**: These devices often produce only a small amount of electricity, which is usually sufficient only for charging small devices like mobile phones or powering low-energy lights.

- **Intermittent Use**: Due to their limited power output, these solutions are best suited for intermittent or emergency use rather than as a primary power source.

Application-Specific Considerations

- **Use Case Matching**: It's essential to match the energy solution to the specific needs and capabilities of the application area. For instance, solar cookers are ideal for sunny regions but may not be practical in areas with frequent cloud cover.

- **Integration with Other Energy Sources**: To overcome limitations, these devices can be used in conjunction with other renewable energy sources or battery storage systems to create a more robust and reliable energy solution.

Embracing these low-tech solutions can significantly enhance energy independence and ensure access to power in emergencies. By combining these tools with modern technology, individuals and communities can achieve a balanced approach to sustainable living and disaster preparedness.

Harnessing the Energy Potential of Organic Waste

Organic waste, including food scraps, agricultural byproducts, and animal manure, represents a significant untapped resource for renewable energy production. Utilizing this waste not only helps in managing waste effectively but also in producing valuable energy. Here's how the energy potential of organic waste can be harnessed:

Understanding Organic Waste as a Resource

Organic waste is a broad category that includes various types of biodegradable material, which can be transformed into valuable energy resources. Recognizing the different types of organic waste and understanding their availability and volume are critical steps in effectively utilizing this waste for energy production. Here's a detailed look at these aspects:

Types of Organic Waste

- **Kitchen Scraps**: This includes leftovers, fruit and vegetable peels, and other food waste typically generated in household and commercial kitchens. The high moisture content and rich organic nature make kitchen scraps excellent candidates for processes like anaerobic digestion.

- **Yard Trimmings**: Grass clippings, leaves, and branches collected from residential and commercial landscaping activities contribute significantly to organic waste. These materials are ideal for composting and can also be used in bioenergy production through processes like gasification.

- **Farm Residues**: Agricultural waste such as crop residues (stalks, husks, seeds), manure, and other byproducts from farming activities are potent sources for energy production. These materials can be used in anaerobic digestion systems to produce biogas or in biomass power plants.

- **Industrial Organic Byproducts**: Waste generated from industries such as food processing, breweries, and paper manufacturing often contains substantial organic components. These byproducts, including whey from cheese production or pulp from juicing, can be processed to extract energy.

Volume and Availability

- **Assessment of Waste Streams**: Conducting a thorough analysis of the organic waste streams available is essential for planning any energy recovery project. This includes identifying the types, quantities, and consistency of waste produced in a given area.

- **Seasonal Variations**: Many types of organic waste, especially agricultural residues, are subject to seasonal variations. Understanding these patterns is crucial for designing systems that can handle fluctuations in waste supply.

- **Geographic Considerations**: The location plays a significant role in determining the availability of organic waste. For example, urban areas may produce large quantities of kitchen scraps and yard trimmings, whereas rural areas might have more access to agricultural residues and manure.

- **Scalability and Economic Viability**: Larger volumes of consistent waste supply are required to make the investment in energy recovery technologies economically viable. Scalable systems that can adjust to the quantity of available waste help optimize both operational efficiency and energy output

Understanding the types and quantities of organic waste available is a foundational step in tapping into this underutilized resource. By effectively assessing and managing these waste streams, communities and businesses can not only reduce environmental impacts but also harness a significant source of renewable energy

Technologies for Energy Recovery from Organic Waste

Organic waste can be converted into valuable energy sources through several technological processes. Each of these technologies offers unique benefits and is suitable for different types of organic materials. Here's a closer look at three key technologies used for energy recovery:

Anaerobic Digestion (AD)

- **Process Description**: Anaerobic digestion is a biological process that occurs in the absence of oxygen, where microorganisms break down organic materials. This process results in the production of biogas, which primarily consists of methane (CH_4) and carbon dioxide (CO_2).

- **Applications**: The biogas produced from AD can be utilized in various ways. It can be burned to generate heat or electricity, or it can be processed further to upgrade the quality to that of natural gas standards, known as biomethane, which can be injected into gas grids or used as vehicle fuel.

- **Benefits**: AD not only helps in reducing the volume of organic waste but also produces digestate, a byproduct that can be used as a nutrient-rich fertilizer, enhancing soil health and reducing the need for chemical fertilizers.

Gasification

- **Process Description**: Gasification involves the conversion of organic materials into syngas (synthesis gas) through high-temperature processing under controlled amounts of oxygen and/or steam. Syngas is a mixture of carbon monoxide, hydrogen, and small amounts of carbon dioxide.

- **Applications**: Syngas produced from gasification can be used directly for heating or power generation. Additionally, it can be converted into liquid fuels or chemicals through further processing, making it a versatile energy carrier.

- **Benefits**: Gasification can process a variety of waste materials, including those not suitable for anaerobic digestion. It typically results in a reduction of waste volume and mass and produces a clean-burning fuel.

Composting

- **Process Description**: Composting is the aerobic decomposition of organic waste, which stabilizes the organic matter. Though primarily used for waste reduction and soil amendment, the process also contributes to energy recovery indirectly by reducing the overall methane emissions from landfill sites.

- **Applications**: The end product of composting is compost, a rich soil conditioner. While compost itself is not used for energy, the process conserves landfill space and reduces greenhouse gas emissions.

- **Benefits**: Composting is a relatively simple and low-cost technology that can be implemented at various scales, from household to industrial. It significantly reduces the need for synthetic fertilizers and promotes sustainable agricultural practices.

These technologies offer robust solutions for managing organic waste while generating energy, promoting a circular economy, and contributing to environmental sustainability. Each method can be tailored to suit specific regional and material conditions, making them versatile tools in waste management and renewable energy sectors.

Implementation Steps for Energy Recovery from Organic Waste

Successfully implementing an energy recovery system from organic waste involves a series of crucial steps. These steps ensure that the system is efficient, sustainable, and tailored to the specific needs and conditions of the area. Here's an overview of the essential stages:

Feasibility Studies

- **Purpose**: Conducting feasibility studies is critical to understand the specific characteristics of the available organic waste, such as its type, quantity, and seasonal availability.

- **Assessment**: Evaluate the local waste generation rates, the types of organic waste available (e.g., food scraps, yard waste, agricultural residues), and their suitability for different energy recovery technologies.

- **Outcome**: Determine the potential energy output, economic viability, and environmental impact of implementing a particular technology.

Technology Selection

- **Criteria**: Choose a suitable technology based on the physical and chemical characteristics of the organic waste, the economic considerations, and the desired end products.

- **Options**: Technologies might include anaerobic digestion for wet organic wastes, gasification for wood and other fibrous organic materials, and composting for mixed organic waste.

- **Local Adaptation**: Consider local environmental regulations, availability of technology providers, and potential markets for the energy and by-products.

System Design

- **Integration**: Design a system that not only fits into the existing waste management infrastructure but also integrates smoothly with local energy systems and markets.

- **Capacity Planning**: The design should match the scale of waste availability and the community's energy needs, which could range from small, community-level systems to large, industrial-scale installations.

- **Flexibility**: Consider future expansion or modification as waste streams and energy demands evolve.

Operational Considerations

- **Waste Preprocessing**: Plan for the necessary preprocessing steps, such as shredding, sorting, and dewatering, depending on the chosen technology and waste material characteristics.

- **System Maintenance**: Develop a maintenance schedule to ensure that all components of the system function efficiently and safely. This includes regular checks and servicing of mechanical and electrical parts.

- **By-product Management**: Establish strategies for managing by-products, such as digestate from anaerobic digestion, which can be used as a high-quality fertilizer. Ensure that these strategies comply with local regulations regarding waste products.

Implementation and Monitoring

- **Pilot Projects**: Before full-scale implementation, conducting pilot projects can help identify potential operational issues and refine the system design.

- **Training and Development**: Train staff and involved parties on the technical, operational, and safety aspects of the system.

- **Performance Monitoring**: Set up a system for ongoing monitoring and evaluation to track the performance of the energy recovery system, optimize operations, and assess environmental and economic impacts.

Implementing a successful energy recovery project from organic waste requires careful planning, a deep understanding of local conditions, and ongoing management to ensure sustainability and effectiveness.

Benefits of Using Organic Waste for Energy

Utilizing organic waste to generate energy offers multiple environmental, economic, and sustainable benefits. Here's a detailed look at the key advantages:

Reduction in Greenhouse Gas Emissions

- **Landfill Impact**: Organic waste in landfills undergoes anaerobic decomposition, producing methane, a greenhouse gas up to 34 times more potent than carbon dioxide over a 100-year period. By diverting organic waste to energy recovery systems, methane emissions are significantly reduced.

- **Sustainable Practice**: Transforming waste into energy prevents the release of large quantities of methane into the atmosphere, contributing to climate change mitigation efforts.

Energy Production

- **Renewable Energy Source**: Through processes like anaerobic digestion and gasification, organic waste is converted into biogas or other forms of renewable energy. This transformation allows for the production of electricity and heat, replacing energy generated from non-renewable sources.

- **Reliability**: Unlike some other forms of renewable energy which might be dependent on weather conditions, the energy from waste processes is stable and can be generated continuously, providing a reliable supply of energy.

Economic Benefits

- **Reduced Disposal Costs**: Processing organic waste for energy generation reduces the amount of waste sent to landfills, thereby lowering landfill usage fees and extending landfill lifespans.

- **Revenue Generation**: Energy produced from organic waste can be sold back to the grid, and by-products like digestate can be used or sold as biofertilizers. This generates additional income streams for businesses or communities.

- **Job Creation**: Developing and maintaining energy recovery facilities and systems creates jobs in engineering, operations, and maintenance, contributing to local economies.

Additional Environmental Benefits

- **Soil Health**: By-products from processes like anaerobic digestion can be refined and used as organic fertilizers, enhancing soil health and reducing dependency on chemical fertilizers.

- **Waste Reduction**: Implementing energy recovery from organic waste contributes to a reduction in the overall volume of waste, easing the burden on municipal waste management systems.

The utilization of organic waste for energy not only helps in combating global warming but also supports a circular economy, turning what would be waste into valuable energy and materials. This approach aligns with global efforts towards sustainability, promoting a cleaner, more resource-efficient future.

Challenges and Barriers to Implementing Organic Waste to Energy Solutions

Implementing systems for converting organic waste to energy involves several challenges and barriers that can affect their widespread adoption. Understanding these obstacles is crucial for effective planning and implementation:

Economic Viability

- **High Initial Investment**: Establishing facilities for energy recovery from organic waste requires significant upfront investment. The cost of technology, infrastructure, and necessary expertise can be prohibitive, especially for smaller or less economically developed areas.

- **Need for Economic Incentives**: Without financial incentives such as subsidies, tax breaks, or grants, the return on investment for these technologies can be unattractive to private investors. Economic viability often relies on supportive financial frameworks to become feasible.

Regulatory Frameworks

- **Complex Legal Landscape**: Regulatory support is essential, but the legal environment surrounding waste-to-energy projects can be complex and varied across jurisdictions. Adequate policies and regulations that support the adoption and integration of these technologies into existing waste management systems are often lacking.

- **Incentive and Support Policies**: There is a need for coherent policies that not only regulate but also actively promote the use of organic waste for energy. This includes renewable energy credits, streamlined permitting processes, and standards for the use and sale of by-products like digestate.

Public Acceptance and Education

- **Community Engagement**: Projects that convert waste to energy can face opposition from local communities due to concerns about potential smells, increased traffic, or emissions. Successful implementation requires active engagement with communities to address these concerns transparently.

- **Educational Initiatives**: There is often a lack of understanding about the benefits of waste to energy projects and the technologies involved. Educational campaigns and public consultations are crucial to improving knowledge and acceptance of these initiatives.

- **Transparency and Trust**: Building trust through transparency about the environmental impacts, economic benefits, and operational aspects of waste to energy facilities can lead to greater community support.

Technical Challenges

- **Technology Fit**: The effectiveness of different technologies varies depending on the type of organic waste and local conditions. Finding the most suitable technology that matches the specific characteristics of the waste available can be a barrier.

- **Maintenance and Operation**: Advanced waste-to-energy technologies require skilled personnel for operation and ongoing maintenance. The shortage of qualified technicians and the need for continuous technical support can hinder the operation of these facilities.

Addressing these challenges involves coordinated efforts between governments, businesses, and communities. It also requires innovative financial strategies, supportive policy frameworks, and strong leadership to ensure that the potential benefits of converting organic waste to energy are fully realized.

Book 8: Housing

Designing and Building Sustainable Shelters

Creating sustainable shelters involves integrating environmentally friendly practices and materials throughout the design and building processes. These shelters are designed to minimize environmental impact, provide energy efficiency, and create a healthy living space for occupants. Here's an in-depth look at the key considerations and methods involved in designing and building sustainable shelters:

Key Principles of Sustainable Shelter Design

Creating sustainable and energy-efficient shelters is pivotal in reducing environmental impact and promoting healthier living spaces. This approach encompasses a range of practices from the strategic orientation of the building to the careful selection of construction materials.

Energy Efficiency in Shelter Design

Energy efficiency is a cornerstone of sustainable architecture. Here's how it can be effectively integrated into the design and construction of shelters:

- **Orientation and Layout**: The first step in designing an energy-efficient shelter involves orienting the structure to maximize natural light and heat, which can significantly reduce reliance on artificial lighting and heating systems. In the northern hemisphere, this typically means having large, south-facing windows that allow for ample sunlight during the winter months while minimizing exposure to the harsh summer sun.

- **Insulation**: Proper insulation is crucial in maintaining internal temperatures, reducing the energy required for heating and cooling. Materials such as spray foam, wool, or recycled denim are effective in preventing heat loss during colder months and keeping interiors cool in warm weather.

- **Passive Solar Heating**: Utilizing passive solar heating involves designing with materials that absorb and slowly release the sun's heat. Elements such as thermal mass floors and walls can store heat during the day and release it at night, maintaining a comfortable temperature inside the shelter.

- **Natural Cooling Strategies**: Incorporating features like strategic window placement, thermal mass, and natural ventilation can help in cooling homes naturally. Overhangs, shaded areas, and green roofs also reduce the need for air conditioning by cooling the environment around the building.

Use of Sustainable Materials

Selecting the right materials is essential for ensuring the sustainability of a shelter:

- **Recycled and Reclaimed Materials**: Using recycled content in construction, such as reclaimed wood, recycled metal, or glass, reduces the demand for virgin materials and the environmental degradation associated with their extraction and processing.

- **Rapidly Renewable Materials**: Materials such as bamboo, cork, and straw are considered rapidly renewable because they regenerate quickly, making them a more sustainable choice than traditional hardwoods.

- **Non-Toxic Materials**: Choosing materials that do not emit harmful chemicals is vital for indoor air quality. Low-VOC (volatile organic compounds) paints, natural fiber insulation, and formaldehyde-free cabinets are examples of healthier choices that contribute to a safer indoor environment.

Water Conservation in Sustainable Shelters

Water conservation is critical in reducing the environmental impact of a building and ensuring sustainability in regions facing water scarcity. Here are key strategies to incorporate:

- **Water-Saving Fixtures**: Installing low-flow faucets, showerheads, and dual-flush or composting toilets can significantly reduce water usage in residential and commercial buildings. These fixtures are designed to provide the necessary functionality while minimizing water waste.

- **Rainwater Harvesting Systems**: Designing a shelter with a rainwater collection system can offset the consumption of municipal water supplies. Collected rainwater can be used for irrigation, flushing toilets, and, with proper treatment, for drinking and other household uses.

- **Greywater Reuse Systems**: Greywater, which includes water from sinks, showers, and washing machines, can be treated and reused for landscape irrigation and other non-potable uses. Integrating a greywater system helps in reducing the fresh water demand substantially.

Waste Reduction Strategies

Minimizing waste during the construction and operational phases of a shelter is crucial for sustainable practices:

- **On-site Management of Construction Waste**: Implementing strategies for waste sorting, recycling, and reuse during construction can dramatically reduce the amount of waste sent to land filed. Planning for material optimization and on-site management reduces leftovers and excess materials.

- **Designing for Flexibility and Future Reuse**: Constructing buildings with adaptable designs allows for easy modifications in the future, thereby extending the building's life and reducing the need for new materials. Using demountable and reusable systems for interior partitions, for example, can facilitate future renovations or repurposing without significant waste.

- **Materials Passport**: Implementing a materials passport approach, where every material used in the construction is documented for its potential reuse and recyclability, can enhance the sustainability of buildings by planning for end-of-life recovery and reuse.

Steps in Building Sustainable Shelters

Creating sustainable shelters begins with thoughtful site selection and detailed planning, followed by an intentional design phase that incorporates key sustainability features. Each step is critical in ensuring that the shelter not only minimizes environmental impact but also enhances the comfort and well-being of its occupants. Here's a deep dive into these foundational stages:

1. Site Selection and Planning

Choosing the right location for constructing a sustainable shelter is pivotal in its overall environmental performance and sustainability. Here are the key considerations:

- **Environmental Impact Minimization**: Select a site that requires minimal alteration, preserving the natural landscape and existing ecosystems as much as possible. This approach helps maintain biodiversity and provides natural amenities to the building occupants.

- **Topography**: Evaluate the site's topography for opportunities to integrate the building into the landscape seamlessly. For example, building into a hillside can provide natural insulation and protection from elements, reducing energy needs for heating and cooling.

- **Vegetation and Ecological Features**: Preserve mature vegetation which can provide natural windbreaks, shade, and cooling. Consider the ecological features of the site to enhance the design, such as incorporating rain gardens or permeable surfaces that contribute to natural water management and reduce runoff.

- **Orientation**: Plan the site layout to take advantage of the sun's path and prevailing winds. Proper orientation can significantly improve the building's energy efficiency by optimizing for solar gain in colder months and natural cooling in warmer months.

2. Design Phase

Once the site is selected, the design phase focuses on developing a blueprint that aligns with sustainability goals:

- **Natural Lighting**: Maximize the use of natural light to reduce reliance on artificial lighting. Thoughtful window placement not only saves energy but also enhances the indoor environmental quality for occupants.

- **Thermal Mass**: Utilize materials with high thermal mass such as concrete or stone in strategic locations. These materials can absorb and store heat from the sun during the day and release it at night, helping to stabilize indoor temperatures.

- **Efficient Spatial Layout**: Design the interior layout to maximize airflow and natural temperature regulation. Open layouts can enhance air circulation, while zoned designs allow for heating or cooling only occupied areas, reducing energy waste.

3. Material Selection

Choosing the right materials is fundamental to the sustainability of the shelter. Here are the key considerations:

- **Lifecycle Environmental Impact**: Evaluate materials based on the entire lifecycle, including the energy consumed in production, transportation, installation, maintenance, and disposal. Opt for materials that offer durability and have minimal environmental impact throughout their lifecycle.

- **Locally Sourced Materials**: Utilize materials that are sourced locally whenever possible. Local sourcing not only supports the local economy but also significantly reduces the carbon footprint associated with long-distance transportation of materials.

- **Sustainable Harvesting and Manufacturing**: Choose materials that are harvested or produced using sustainable methods. For instance, certified wood from responsibly managed forests, recycled metal, or bricks made from local clay can be excellent sustainable choices.

- **Energy-Efficient Materials**: Incorporate materials that contribute to energy efficiency, such as high-performance windows, insulated siding, and roofing materials that enhance thermal efficiency.

- **Non-Toxic and Natural Materials**: Select materials that do not emit harmful chemicals or volatile organic compounds (VOCs). Using natural or minimally processed materials can improve indoor air quality and reduce health risks for occupants.

4. Construction Practices

The methods used during the construction process play a significant role in the environmental footprint of the building project:

- **Minimizing On-Site Damage**: Carefully plan site preparation and construction to avoid unnecessary damage to the surrounding environment. Techniques such as preserving topsoil, controlling erosion, and protecting nearby water sources are essential.

- **Efficient Machinery and Tools**: Utilize energy-efficient and properly maintained construction machinery to reduce emissions and energy use. Where possible, electric-powered equipment can be a more sustainable option compared to diesel or gas-powered machinery.

- **Recycling Construction Debris**: Implement systems to recycle or reuse construction debris. Materials like wood, metal, and concrete can often be recycled, reducing waste and conserving resources.

- **Sustainable Building Techniques**: Employ construction techniques that reduce energy use and material waste, such as modular construction or prefabricated components that can be assembled on-site. These methods not only speed up the construction process but also increase precision and reduce waste.

- **Engaging Skilled Workers**: Work with contractors and workers who have experience and training in sustainable construction practices. Their expertise can ensure that sustainable methods are effectively implemented throughout the building process.

5. Renewable Energy Integration

Incorporating renewable energy systems during the initial phases of construction can have substantial benefits for both the environment and the homeowner:

- **Solar Energy**: Installing solar panels on rooftops or in open areas of the property can significantly reduce reliance on non-renewable energy sources. These systems can provide electricity for heating, cooling, and running household appliances.

- **Wind Energy**: In suitable locations, small wind turbines can supplement home energy needs. This is particularly effective in rural or open areas where wind speeds are favorable.

- **Geothermal Heating and Cooling**: Implementing geothermal systems that utilize the stable temperatures underground to heat and cool the home can offer a highly efficient alternative to traditional HVAC systems.

- **Hybrid Systems**: Combining two or more renewable energy sources can ensure a more constant and reliable supply of energy, thereby enhancing energy security and further reducing environmental impacts.

6. Landscape and External Environment

The design of the external environment is crucial in promoting ecological health and sustainability:

- **Native Plant Landscaping**: Using native plants in garden and landscape design not only reduces the need for irrigation and chemical fertilizers but also supports local wildlife, including pollinators and birds.

- **Permeable Surfaces**: Incorporating permeable paving materials in driveways, walkways, and patios allows rainwater to infiltrate into the ground, reducing runoff and preventing erosion.

- **Water Management Features**: Design features like rain gardens, swales, and retention ponds can effectively manage stormwater on-site, preventing pollution of nearby waterways and reducing the burden on municipal stormwater systems.

- **Integration with Local Ecosystems**: Designing outdoor spaces to blend seamlessly with the surrounding environment helps preserve local biodiversity and ecological balance. This includes the strategic placement of trees and shrubs to provide natural windbreaks and enhance privacy, while also offering habitat for wildlife.

- **Outdoor Living Spaces**: Designing outdoor areas for human use, such as patios, decks, and gardens, encourages interaction with the natural environment, promoting a healthier lifestyle and greater appreciation for the outdoors.

Building sustainable shelters is a comprehensive approach that requires thoughtful design, careful material selection, and innovative construction techniques. These structures not only reduce the impact on the environment but also offer long-term economic benefits through reduced operating costs, proving that sustainability is both a practical and an ethical choice in modern architecture.

Effective Insulation and Heating in Sustainable Housing

In the realm of sustainable building practices, effective insulation and heating systems play pivotal roles in enhancing energy efficiency and ensuring comfortable living conditions. Here's a detailed look at how proper insulation and efficient heating solutions contribute to sustainable housing:

Insulation Techniques for Sustainable Housing

Effective insulation is a cornerstone of energy-efficient and sustainable housing. By properly insulating a home, you can significantly minimize heat loss during the winter and reduce heat gain during the summer, leading to substantial energy savings and a smaller carbon footprint. Here is a detailed overview of the key techniques and considerations in home insulation:

Material Selection

Choosing the right insulation material is essential for achieving optimal thermal efficiency. Various insulation materials cater to different needs depending on the climate and specific areas of application within the building:

- **Natural Fibers**: Wool and cotton are excellent for those seeking sustainable and less toxic options. These materials are biodegradable and have good thermal properties, making them suitable for residential use.

- **Recycled Materials**: Cellulose insulation, made from recycled paper products, is both eco-friendly and effective. It has a high R-value and can easily be installed in various spaces, making it ideal for retrofitting older homes.

- **Synthetic Materials**: Polystyrene and polyurethane foams are known for their high R-values and moisture resistance. These foams can be applied as sprays or rigid panels and are particularly effective in preventing heat transfer in extreme climates.

Thermal Performance

- **R-values**: Insulation's effectiveness is quantified by its R-value, which measures its ability to resist heat flow. Materials with higher R-values are more effective insulators. When selecting insulation, it's important to choose materials that meet or exceed local building codes for R-value, ensuring optimal energy efficiency.

Comprehensive Coverage

- **Addressing All Building Parts**: Effective insulation involves more than just filling the walls. It includes:

    o **Walls**: Insulating both exterior and interior walls can dramatically reduce heating and cooling costs.

    o **Roofs and Attics**: These areas are crucial for preventing heat loss in the winter and heat gain in the summer. Adequate insulation here can improve the overall energy performance of a home.

    o **Floors and Foundations**: Insulating these areas helps maintain a stable indoor temperature and reduces the risk of moisture-related issues.

- **Minimizing Thermal Bridges**: Thermal bridging occurs when materials that conduct heat bypass the insulation, such as along metal studs or uninsulated window frames. Addressing these bridges by using thermal break materials or increasing insulation around these areas is essential for maintaining an energy-efficient building envelope.

Implementation Tips

- **Professional Assessment**: Consider hiring a professional to assess existing insulation and identify areas where improvements are needed. This can include thermal imaging to visually identify areas of poor insulation.

- **Air Sealing**: Complement your insulation efforts with thorough air sealing, especially around doors, windows, and where utility lines enter the house, to prevent warm air from escaping and cold air from entering.

- **Ventilation**: Ensure that your home maintains adequate ventilation to prevent indoor air quality issues. This is particularly important in tightly sealed and well-insulated homes.

Sustainable Heating Systems for Energy-Efficient Homes

Sustainable heating systems are crucial for reducing our dependence on non-renewable energy sources and enhancing energy efficiency. When integrated during the design phase of a building, these systems can optimize both performance and energy savings, leading to long-term cost benefits and environmental impact reduction. Here's an overview of the most effective sustainable heating systems:

Solar Heating

Solar heating systems harness solar energy to directly heat air or water, going beyond just electricity generation:

- **Solar Thermal Panels**: These panels absorb sunlight and convert it into heat, which is then transferred to air or water circulation systems within the home.
- **Benefits**: Solar heating can significantly reduce dependency on conventional heating methods like gas or electric furnaces, cutting down on energy costs and greenhouse gas emissions.

Heat Pumps

Heat pumps are a versatile and efficient solution for heating and cooling:

- **Types**: There are several types of heat pumps, including air-source, ground-source (geothermal), and water-source, each using the ambient temperature of their respective source to regulate the temperature inside the home.
- **Efficiency**: They operate by extracting heat from natural sources (air, ground, water) and pumping it indoors. During warmer months, the process can be reversed to cool the interiors, making heat pumps an all-in-one climate control solution.

Pellet Stoves and Biomass Boilers

These systems use renewable materials as their fuel source:

- **Pellet Stoves**: Compact devices that burn compressed wood or biomass pellets to create heat. They are ideal for heating smaller spaces or used in conjunction with other heating systems.
- **Biomach Boilers**: Larger systems that can heat entire buildings by burning wood chips, pellets, or other biomass materials.
- **Sustainability and Efficiency**: Both systems produce fewer emissions compared to traditional wood stoves and are capable of achieving higher efficiency with controlled burning technology.

Radiant Floor Heating

This system offers an even distribution of heat and enhances comfort without the noise and drafts of traditional forced-air systems:

- **Installation**: Heating elements are installed beneath the floor surface, and as they heat up, the warmth is radiated upwards into the room.
- **Benefits**: Radiant heating provides uniform warmth from the floor up, allowing lower thermostat settings while maintaining comfort. It's particularly efficient in homes with good insulation, as it minimizes heat loss that typically occurs with conventional heating systems.

Considerations for Implementation

- **Integration with Design**: Integrating these heating systems during the initial design phase of a building is critical for achieving optimal energy efficiency. It allows for the architectural integration of components like solar panels and ensures that the landscape can accommodate elements like ground-source heat pumps.
- **Cost vs. Benefit**: While the initial setup costs can be high, the long-term benefits—reduced energy bills and lower environmental impact—justify the investment.

- **Local Climate and Environment**: The effectiveness of each system varies with local conditions. For example, solar heating is more viable in regions with high sun exposure, while heat pumps are more effective in areas with moderate climates.

Integration of Sustainable Design Principles in Building Architecture

Integrating sustainable design principles into building architecture not only enhances energy efficiency but also reduces long-term operational costs, making it a crucial aspect of modern construction. Here's how architects and builders can incorporate these principles effectively:

Orientation and Layout

The orientation and layout of a building play a pivotal role in leveraging natural resources for heating and cooling:

- **Solar Gain**: By orienting a building to maximize exposure to the south (in the Northern Hemisphere), it can capture optimal solar heat during the winter, reducing the need for artificial heating.

- **Shading and Overhangs**: Strategic placement of overhangs and landscaping can provide shade during the hotter months, thereby reducing cooling needs.

Window Placement and Glazing

Windows are critical components in managing a building's heat balance:

- **Double or Triple-Glazed Windows**: These windows consist of multiple layers of glass with air or gas-filled spaces between them, which significantly improves their insulative properties.

- **Low-Emissivity Coatings**: These coatings are applied to windows to reflect infrared light, keeping heat inside during the winter and outside during the summer.

- **Strategic Placement**: Placing windows on specific sides of the building can harness sunlight in the winter and minimize heat penetration during summer, complementing the natural climate control.

Sealing and Ventilation

Creating an airtight building envelope is key to maintaining energy efficiency, but it must be balanced with the need for indoor air quality:

- **Airtight Sealing**: Ensuring that all joints, seams, and openings are sealed prevents unwanted air leaks, which can lead to heat loss and gain.

- **Ventilation Systems**: Integrating advanced ventilation systems like Heat Recovery Ventilators (HRV) or Energy Recovery Ventilators (ERV) helps maintain air quality without sacrificing thermal comfort. These systems recover heat from exhaust air and transfer it to incoming fresh air, reducing the energy required to heat new air entering the building

Conclusion

Incorporating effective insulation and efficient heating systems in the design and construction of sustainable shelters is crucial for reducing energy consumption, enhancing comfort, and minimizing environmental impact. These technologies not only support sustainable development goals but also offer long-term economic benefits through reduced energy costs. Regular maintenance and monitoring further ensure these systems operate at their optimal capacity, sustaining their benefits over the shelter's lifespan.

Sustainable Waste Management in Housing

Sustainable waste management is a critical component of modern housing design, emphasizing the reduction of waste production, enhancement of recycling efforts, and responsible disposal practices. This approach not only minimizes environmental impact but also promotes resource efficiency and sustainability in residential developments.

Principles of Sustainable Waste Management

Sustainable waste management is essential for reducing environmental impacts and promoting efficiency within residential developments. The following principles outline key strategies for incorporating sustainable waste management practices into housing projects:

Waste Minimization

Waste minimization focuses on reducing the amount of waste generated from the outset. By implementing efficient design and construction techniques, developers and builders can significantly decrease waste production. Key methods include:

- **Precise Material Ordering**: Calculating exact material requirements to prevent excess.

- **Use of Prefabricated Elements**: Employing prefabricated parts in construction to reduce cut-offs and waste on-site.

- **Selection of Durable Materials**: Opting for materials that are robust and have longer life spans, which minimizes the need for replacements and reduces waste.

Recycling and Reuse

Recycling and reuse are critical components of sustainable waste management. These practices not only help conserve resources but also decrease the environmental burden associated with producing new materials. Effective strategies include:

- **Separation Systems**: Installing well-marked and conveniently located recycling bins to ensure effective separation of waste types, which facilitates higher-quality recycling.

- **Reusing Materials**: Encouraging the salvage and reuse of materials during construction and renovation, which can drastically cut the need for new resources and lower overall project costs.

Composting

Composting is a valuable waste management tool, particularly in residential areas with available outdoor space:

- **Organic Waste Transformation**: By composting kitchen scraps and yard waste, households and communities can convert organic waste into nutrient-rich compost.

- **Soil Quality Improvement**: The use of compost in gardens and landscaping enriches the soil, improving its structure and fertility, and reduces the reliance on chemical fertilizers.

- **Community Engagement**: Implementing community composting initiatives can engage residents, increase awareness of sustainability issues, and provide direct benefits through improved garden productivity.

Implementation Strategies for Sustainable Waste Management

To effectively implement sustainable waste management in residential areas, a combination of thoughtful design, infrastructure development, and community engagement is required. The following strategies outline how developers, local governments, and community leaders can enhance waste management practices:

Design for Disassembly

This approach focuses on constructing buildings that can be easily taken apart at the end of their lifecycle, facilitating the reuse and recycling of materials. Key aspects include:

- **Use of Mechanical Fasteners**: Employing screws and bolts instead of permanent adhesives or welding. This makes it easier to remove and reuse components without damaging them.

- **Modular Construction**: Designing buildings in modules or sections that can be independently disassembled and reused. This not only aids in waste reduction but also allows for flexibility in use during the building's lifecycle.

Resource Recovery Facilities

Incorporating resource recovery facilities within community infrastructure can greatly enhance the efficiency of waste management systems. These facilities serve multiple functions:

- **Centralized Recycling**: Providing a centralized location where all recyclables can be processed, which improves the efficiency and effectiveness of recycling programs.

- **Organic Waste Management**: Facilitating the composting of organic materials, which reduces landfill use and produces valuable compost for local agricultural or landscaping use.

- **E-Waste Handling**: Safely processing electronic waste, which often contains hazardous materials and valuable resources that can be recovered and reused.

Education and Awareness

Educating residents about sustainable waste practices is crucial for the success of any waste management strategy. Effective education and awareness efforts include:

- **Informational Campaigns**: Distributing brochures, flyers, and emails that provide details on how to participate in and benefit from recycling and waste reduction efforts.

- **Workshops and Seminars**: Organizing regular events that educate the public on the importance of waste segregation, the benefits of recycling, and techniques for effective composting.

- **Online Resources**: Developing digital platforms or apps that provide easy access to information about waste management services, collection schedules, and guidelines for proper waste disposal.

Innovative Technologies in Waste Management

Advancements in technology offer remarkable opportunities to enhance waste management practices, particularly in residential and construction settings. Here are some key innovations that are transforming the way communities handle waste:

Smart Waste Management Systems

Smart technology is revolutionizing residential waste management by making it more efficient and environmentally friendly:

- **Smart Bins**: These are equipped with sensors that monitor waste levels and can even determine the composition of the waste. They are connected to a central system that analyzes the data to optimize waste collection routes and schedules, reducing unnecessary pickups and saving fuel.

- **Automated Waste Sorting**: Utilizes advanced sorting technologies that can separate recyclables from general waste with high precision. This not only improves recycling rates but also reduces contamination in the recycling stream.

Construction Waste Management Plans

Implementing a formalized plan for managing waste during construction and demolition phases is crucial for reducing the environmental impact of building activities:

- **Pre-construction Waste Assessment**: Before construction begins, an assessment is conducted to identify the types and quantities of materials that will be generated and to plan for their reduction, reuse, or recycling.

- **Waste Reduction Targets**: Setting clear targets for waste reduction and recycling as part of the construction waste management plan. These targets hold contractors and developers accountable and encourage the adoption of sustainable practices.

- **Documentation and Tracking**: Maintaining detailed records of waste generated and how it is processed. This documentation is often required to comply with local regulations and can help in obtaining green building certifications.

Benefits and Impact

The adoption of these innovative technologies and strategies has several positive outcomes:

- **Efficiency and Cost Reduction**: Smart systems reduce operational costs by optimizing waste collection routes and schedules, thus lowering the overall carbon footprint associated with waste management.

- **Regulatory Compliance**: Construction waste management plans ensure compliance with local and international waste disposal regulations, reducing the risk of fines and facilitating sustainable development.

- **Environmental Sustainability**: By increasing the efficiency of waste collection and enhancing recycling processes, these technologies help reduce landfill use, lower greenhouse gas emissions, and conserve natural resources.

Together, smart waste management systems and construction waste management plans represent a shift towards more sustainable and responsible waste handling, essential for building resilient, environmentally conscious communities.

Conclusion

Sustainable waste management in housing is not just about handling waste responsibly but also about redesigning the lifecycle of materials used in the construction and maintenance of homes. By embedding sustainability into every stage of a building's lifecycle, from design and construction to demolition and disposal, we can significantly reduce the environmental impact of our living spaces and promote a more sustainable future.

Book 9: Emergency Preparedness

Essential Survival Kits

In emergency situations, having a well-prepared survival kit can be crucial for ensuring safety and resilience. A comprehensive survival kit should be tailored to meet the specific needs of the environment and potential scenarios one might face. Here's a guide to assembling an essential survival kit:

Components of an Essential Survival First Aid Kit

1. Water and Hydration Supplies

Water is the most critical component in any survival kit, as dehydration can become a serious threat within hours under some conditions.

- **Bottled Water**: Include commercially bottled water, which is safe for drinking and has a longer shelf life. Ensure you have at least one gallon of water per person per day, accounting for both drinking and sanitary needs. For a family of four, for example, you would need twelve gallons for a three-day period.

- **Water Purification Methods**: In cases where bottled water runs out or isn't available, having water purification tools is essential. Portable water filters, iodine tablets, or chlorine drops can purify water from natural sources, making it safe to drink. A portable water filter can remove pathogens and contaminants from water sources like streams or lakes. Iodine tablets are lightweight and effective but should be used according to instructions as they can leave an aftertaste.

2. Food Supplies

Having a reliable supply of food can help maintain strength and morale during an emergency.

- **Non-perishable Foods**: Choose foods that do not require refrigeration, cooking, or special preparation. Examples include:

    - **Canned Goods**: Fruits, vegetables, beans, and meats. Ensure you have a manual can opener or choose cans with pull-tabs.

    - **Ready-to-Eat Meals**: Military-style MREs (Meals Ready to Eat) provide a balanced diet and are designed to last for years.

    - **Energy Bars and Granola Bars**: These are calorie-dense and can provide quick energy boosts.

    - **Dried Fruits and Nuts**: Offer nutritional value and long shelf life, and they are light to carry.

- **Storage**: Pack these items in airtight, pest-resistant containers to prevent spoilage and extend their shelf life. Consider vacuum-sealed bags to save space and protect food from moisture and pests.

- **Supply**: Ensure at least a three-day supply per person as a minimum. This amount could be higher depending on the specific circumstances, such as the isolation of your location or known risks that could extend the duration of an emergency.

Planning and Rotation

- **Regular Review**: Every six to twelve months, review your food and water supplies for any expired or near-expiration items and replace them as necessary. This helps ensure that everything in your survival kit is fresh and effective when needed.

- **Dietary Considerations**: Remember to consider any special dietary needs or restrictions your family members might have, such as allergies, baby food requirements, or medical conditions influenced by diet.

3. First Aid Kit

A well-stocked first aid kit is essential for handling medical emergencies during any crisis, helping to manage injuries and prevent infections until professional medical help can be accessed.

- **Basic Supplies**: Include various sizes of sterile bandages, adhesive bandages, gauze pads, and adhesive tape for dressing wounds. Antiseptic wipes and antibiotic ointments are crucial for cleaning cuts and preventing infections.

- **Medications**: Stock up on pain relievers such as acetaminophen or ibuprofen, anti-inflammatory drugs, antihistamines for allergic reactions, and any prescribed medications that family members may need regularly. Include medications for stomach upset and diarrhea.

- **Tools and Equipment**: Scissors, tweezers, and safety pins are useful for cutting bandages or removing splinters. A thermometer can monitor potential fevers, while disposable gloves can prevent contamination.

- **First Aid Manual**: Include a comprehensive first aid guide that covers a variety of emergency situations. This manual should provide instructions on how to handle common injuries, such as sprains, fractures, burns, and CPR procedures.

- **Special Needs**: Consider the specific health needs of your family; for example, if someone has severe allergies, include an epinephrine injector (EpiPen). Asthma inhalers and glucose monitoring devices are also important for those with chronic conditions.

4. Clothing and Bedding

In an emergency, proper clothing and bedding are vital for maintaining body temperature and protection from the elements.

- **Layered Clothing**: Pack clothing that can be layered to adjust to varying temperatures. Moisture-wicking fabrics are ideal for the base layer, insulating materials such as fleece or wool for the middle layer, and waterproof, breathable garments for the outer layer.

- **Footwear**: Include sturdy, comfortable footwear that is suitable for potentially rugged or unstable terrain. Waterproof boots are particularly important in wet conditions.

- **Rain Gear**: A durable raincoat, waterproof pants, and waterproof covers for bags and other equipment can protect against wet weather, helping to prevent hypothermia in cold conditions.

- **Thermal Blankets and Sleeping Bags**: Pack lightweight but effective thermal blankets and a sleeping bag per person. These items should be rated for temperatures lower than the coldest expected weather to ensure comfort and safety.

- **Extras**: Hats, gloves, and thermal socks are small but critical items for maintaining body heat in cold climates. Sun hats and UV-protective clothing are just as important for hot and sunny conditions.

Storage and Accessibility

- **Packaging**: Pack these items in waterproof bags or containers to protect them from moisture and other elements. Clearly label each segment of your kit to allow quick access in an emergency.

- **Maintenance**: Regularly check the condition of all packed items, especially medications and batteries, and replace them as needed to ensure they remain functional when needed.

5. Lighting and Power

Effective lighting and power sources are crucial for safety and functionality during power outages or when navigating environments without natural light.

- **Flashlights**: Include several high-quality, durable flashlights. LED flashlights are preferred for their long battery life and bright light output. Consider headlamps as well for hands-free operation which is essential when performing tasks at night or in dark places.

- **Extra Batteries**: Stock a sufficient supply of batteries, ensuring they are the correct size for all devices that require them. Rotate batteries periodically and check for expiration dates to maintain functionality when needed.

- **Solar-Powered Chargers**: These allow for the charging of small devices like cell phones and GPS units without reliance on the electrical grid. They are particularly useful during extended periods of power outage.

- **Hand-Crank Chargers**: Hand-crank chargers are invaluable in emergency kits because they provide power on demand, regardless of weather conditions or time of day, ensuring you always have a way to recharge essential devices.

- **Light Sticks**: Chemical light sticks are a reliable source of light in emergency situations. They can be used to mark locations, signal for help, or provide light during a power outage without consuming any battery power.

6. Communication Tools

Maintaining the ability to receive updates and communicate with others is critical during an emergency.

- **Battery-Powered or Hand-Crank Radio**: This is essential for receiving weather alerts and emergency broadcasts. Models that feature NOAA (National Oceanic and Atmospheric Administration) channels are recommended as they provide continuous updates specific to your area.

- **Emergency Cell Phone**: Keep a dedicated, fully charged cell phone in your kit, preferably with prepaid minutes or a long-lasting battery. This phone should have emergency numbers pre-stored and be kept in a waterproof bag to protect it from damage.

- **Backup Battery or Solar Charger**: Include a backup power source for your emergency cell phone and other essential communication devices. A solar charger can be particularly useful if the power outage lasts for an extended period.

- **Two-Way Radios**: If you're in an area with poor cell service or if you need to communicate with a group during an evacuation or while managing an emergency, two-way radios can be invaluable.

Storage and Care

- **Waterproof Containers**: Store all lighting and communication devices in waterproof and shockproof containers to protect them from moisture and damage.

- **Regular Checks**: Regularly test all devices to ensure they are working correctly. Charge solar chargers and hand-crank devices every few months and after every use to keep them in good working condition.

- **Accessibility**: Keep these items where they can be easily accessed in the dark or in a rush. It's often useful to keep a flashlight near your bed and another with your emergency kit.

7. Tools and Supplies

Having a set of versatile tools and supplies is essential for handling minor repairs, navigating, and ensuring safety in emergency situations.

- **Multi-Tool or Swiss Army Knife**: These compact tools combine several functions in one, including knives, screwdrivers, can openers, and scissors. They are invaluable for making quick repairs, opening cans, cutting ropes, or performing a variety of other tasks that may be needed in an emergency.

- **Duct Tape**: Known for its strength and versatility, duct tape can be used to repair tears in clothing or tarps, seal containers, make emergency bandages, or even temporarily fix broken windows or equipment.

- **Scissors**: Useful for cutting bandages, clothes, or other materials in a medical emergency or for general use.

- **Plastic Ties**: Can be used to secure tarps, make temporary repairs, or organize wires and cords to prevent tangling and mishaps.

- **Whistle**: A whistle is a simple, yet effective tool for signaling for help. It's loud enough to be heard at a distance and doesn't require any power or batteries to operate.

- **Local Maps and Navigation Tools**: Even if GPS is available, having local maps can be invaluable, especially if electronic navigation tools fail. Include a compass to aid in orienting yourself if you need to navigate without landmarks.

8. Personal Hygiene Items

Maintaining personal hygiene is crucial for comfort, morale, and health, particularly in prolonged emergencies.

- **Soap and Hand Sanitizer**: Keeping hands clean is crucial to prevent the spread of germs, especially in a situation where access to clean water might be limited.

- **Toothbrush and Toothpaste**: Oral hygiene is important to maintain overall health and prevent infections.

- **Sanitary Napkins and/or Tampons**: Essential for women during emergencies, providing comfort and maintaining hygiene.

- **Toilet Paper**: Include a roll or two of toilet phone.paper; if space allows, this can be essential for maintaining sanitation.

- **Additional Considerations for Family Members**:

  o **For Babies**: Include diapers, baby wipes, diaper rash cream, and a bottle if applicable.

  o **For Elderly or Those with Medical Conditions**: Consider any special hygiene products they might need, such as denture care items or incontinence pads.

Storage and Accessibility

- **Waterproof Bags**: Keep all hygiene items in waterproof containers or zip-lock bags to protect them from moisture and contamination.

- **Easily Accessible**: Organize tools and hygiene items so they can be quickly and easily accessed without having to dig through other supplies.

9. Important Documents

Securing important documents is crucial in emergencies, especially when you might need to quickly establish your identity, access resources, or manage your affairs during or after a disaster.

- **Types of Documents**: Include personal identification (IDs, passports), financial documents (bank account information, credit card details), insurance policies (health, home, vehicle), and legal documents (wills, deeds). Also, include family records such as marriage certificates and birth certificates.

- **Emergency Contact Information**: Have a list of emergency contacts, including family members, close friends, doctors, and any relevant professional contacts like lawyers or insurance agents.

- **Medical Information**: Copies of medical prescriptions are vital if medication needs to be refilled after a disaster. Also, include information about any medical conditions or allergies that could affect the treatment during emergencies.

- **Storage**: Keep these documents in a waterproof and fireproof container that is portable enough to take with you during an evacuation. Consider keeping digital copies stored securely online or in a USB flash drive as part of your emergency kit.

10. Specialty Items

Tailoring your emergency kit to include items specific to your environment and potential local risks can greatly enhance your preparedness.

- **Location-Specific Items**:
  - o **For areas prone to snake bites**: Include a snake bite kit with suction devices, bandages, and a lymphatic constrictor.
  - o **For water-based locations**: Water flotation devices or personal flotation devices (PFDs) can be life-saving in floods or accidents near or on bodies of water.
  - o **In mountainous or snow-prone areas**: Carry avalanche transceivers, probes, and snow shovels to help locate and rescue in the case of an avalanche.

- **Personal Items**:
  - o **Eyewear**: If you rely on glasses or contact lenses, include an extra pair and related supplies (like lens solution).
  - o **Children's Items**: If you have children, include items to meet their specific needs, such as diapers, formula, favorite snacks, and comfort items like a stuffed animal.
  - o **Pet Supplies**: Don't forget about the needs of pets—include food, water, a leash, and any medications.

Storage and Accessibility

- **Customization**: Customize the contents of your kit based on a thorough assessment of your family's specific needs and the environmental and seasonal challenges of your area.

- **Accessibility**: Keep specialty items in accessible locations known to all family members. Regularly review and update these items to ensure they are functional and relevant to current needs.

Training and Familiarity

Preparing for emergencies involves not just having the right tools and supplies but also ensuring that every member of the household is knowledgeable and comfortable using them. Training and familiarity are crucial for efficient and effective response when a crisis strikes.

Family Preparedness

- **Knowledge of Kit Contents**: It's important that every family member knows exactly what is in the emergency kit and where it is stored. This includes understanding the specific uses of each item, from first aid supplies to food and water sources.

- **Usage Training**: Practical training on how to use the items in the emergency kit, such as how to purify water using iodine tablets or a portable filter, how to apply a tourniquet, or how to set up and use a hand-crank radio or flashlight. This ensures everyone can utilize the kit's contents under stress.

- **Emergency Drills**: Regularly conduct family drills to practice what to do in case of different types of emergencies (e.g., fires, earthquakes, floods). Drills should include practicing evacuation routes, using communication tools, and accessing and using supplies from the emergency kit.

- **Special Needs Considerations**: Make sure to tailor training and practice sessions to include preparations for family members with special needs, young children, or elderly members who may require additional assistance.

Community Resources

- **Local Emergency Services**: Familiarize yourself with the local emergency response plans and services. Know the locations of shelters, medical facilities, and points of contact within local emergency management agencies.

- **Community Preparedness Programs**: Engage with community emergency preparedness programs if available. These programs offer valuable training and resources, and participating can provide additional insights into local hazards and the community's response strategies.

- **Neighborhood Support Networks**: Establish or join a neighborhood support network. This can be invaluable in a disaster, as communities that work together can more effectively manage local resources and provide assistance to each other.

- **Information Channels**: Stay informed about potential emergencies through reliable sources such as local news outlets, official social media channels, or direct alerts from local government and emergency services. Ensure you understand how to receive and interpret emergency alerts and updates.

Regular Updates and Reviews

- **Kit Inspection and Update**: Regularly review and update the emergency kit. Check expiration dates on food, water, medicine, and batteries, replacing as necessary. Adjust the contents of the kit as family needs and circumstances change, such as changes in medical needs or the addition of new family members (including pets).

- **Plan Revisions**: As you participate in community programs and drills, take the opportunity to refine your family's emergency plan based on new information or changes in local infrastructure and resources.

By emphasizing training and familiarity, not only can you ensure your family is prepared to use the emergency kit effectively, but you also strengthen your overall resilience and capacity to handle potential disasters. Regular engagement with community resources further enhances this preparedness, creating a network of support that can significantly impact your ability to respond effectively in times of need.

Evacuation Plans and Safety Strategies

When disaster strikes, having a clear and practiced evacuation plan is essential for ensuring the safety and security of all individuals involved. A comprehensive evacuation strategy accounts for various types of emergencies, anticipates challenges, and outlines clear steps for safe evacuation. Here are the key components of effective evacuation plans and safety strategies:

Developing an Evacuation Plan

Identifying Hazards

Understanding the specific hazards that your region faces is a crucial first step in preparing an effective evacuation plan. Each type of disaster demands a tailored approach based on its nature and potential impact.

- **Regional Risk Assessment**: Research and acknowledge the natural and man-made disasters most prevalent in your area. For instance, coastal regions might prepare for hurricanes, while seismic zones need plans for earthquakes.

- **Historical Data**: Review historical data on past disasters in your area to understand their frequency, severity, and the areas most affected. This information can be obtained from local government offices or disaster response agencies.

- **Consult Experts**: Engage with local authorities, disaster preparedness experts, and meteorological services to get detailed and accurate information about potential hazards. These experts can offer insights that refine your understanding and preparedness measures.

- **Community Resources**: Participate in community meetings that discuss regional hazards and emergency preparedness. These forums can provide valuable insights and foster a collaborative approach to disaster readiness.

Planning Escape Routes

A well-thought-out plan for evacuation routes is essential to ensure that you can leave quickly and safely during an emergency. This involves mapping out and familiarizing yourself with multiple routes.

- **Primary and Secondary Routes**: Identify the main routes and several alternative routes that you can take to evacuate the area. This is important in case the primary route is obstructed or deemed unsafe during a disaster.

- **Accessibility and Safety**: Check that these routes are accessible to all members of your household, including those with disabilities. Ensure that the routes avoid hazard-prone areas such as flood zones or areas likely to experience heavy debris.

- **Route Maps**: Create detailed maps of these routes and distribute copies to all household members. Consider including landmarks and safe spots where family members can regroup if separated during an evacuation.

- **Practice Drills**: Regularly practice evacuation using these routes to ensure everyone knows them well and can follow them under stress. During drills, evaluate the practicality of the routes and make adjustments as needed.

- **Local Guidance**: Stay informed about the routes recommended by local emergency management officials. Authorities often provide marked evacuation routes and instructions during widespread emergencies, which should be integrated into your personal evacuation plan.

Establishing Meeting Points

In an emergency evacuation, having predetermined meeting points can greatly enhance the safety and coordination of all involved parties. This planning ensures that even if family members are separated during the crisis, they have a familiar and agreed-upon location to regroup.

- **Local Meeting Points**: Choose accessible locations within your neighborhood that are safe from the anticipated disasters, such as community centers, parks, or a neighbor's house that is easy to reach on foot.

- **External Meeting Points**: Identify locations outside of your immediate area, ideally in different directions to account for varying emergency scenarios. These could be relatives' homes, schools, or municipal buildings in nearby towns.

- **Visibility and Accessibility**: Ensure that all chosen meeting points are well-known and easily accessible for all family members, including children and elderly relatives. Consider accessibility for those with disabilities.

- **Practice Visits**: Regularly visit and practice traveling to these meeting points with your family. During these drills, discuss various scenarios in which you might need to use each location.

Organizing Transportation Means

Effective transportation is key to a successful evacuation. Planning involves ensuring that you can leave quickly and safely, considering all family members, including pets.

- **Vehicle Readiness**: Keep a designated emergency vehicle if possible, with at least half a tank of gas at all times and basic maintenance checks up-to-date. Include emergency supplies in the vehicle such as water, snacks, and a first aid kit.

- **Public Transportation**: Familiarize yourself with available public transport options and their operational routes, especially those that lead to your meeting points or out of the hazard zones. Keep a map and schedule of these services.

- **Special Considerations**: Plan for the transportation needs of pets, and any family members with mobility impairments. This might mean arranging for special vehicles or equipment.

Developing a Communication Plan

Communication during an evacuation can be challenging but is crucial for ensuring everyone's safety and for coordinating with emergency services.

- **Contact Lists**: Prepare a comprehensive list of contacts including family members, neighbors, emergency services, and important medical contacts. Distribute this list among all family members and keep a copy in your emergency kit.

- **Communication Methods**: Establish a primary and a backup communication method. Mobile phones with emergency charging options, such as solar chargers or battery packs, are common; however, also consider alternative methods like satellite phones or two-way radios in case of network failure.

- **Check-In Strategy**: Designate a family member or friend outside the immediate area as a central contact point to relay messages and coordinate information among separated family members.

- **Emergency Alerts**: Subscribe to local emergency notification services that provide real-time alerts via SMS or email about looming hazards. Ensure all family members are included in these subscriptions.

Safety Strategies During Evacuation

Taking Early Action

The promptness of your response to an evacuation order can significantly influence your safety during emergencies. Here's what to consider:

- **Immediate Departure**: As soon as evacuation is advised, leave without delay. Early evacuation helps avoid the rush that can occur if everyone leaves simultaneously when the danger intensifies.

- **Stay Informed**: Continuously monitor reliable news sources, weather updates, and official communications. Understanding the nature of the threat can help you decide the urgency of your departure.

- **Understand the Signals**: Know the different alarms and signals used in your area for different types of emergencies, so you can respond appropriately to each.

Managing Essential Documents and Kits

In the chaos of an evacuation, having all necessary supplies and documents organized can provide both practical benefits and peace of mind.

- **Emergency Kits**: Each family member should have an emergency kit tailored to their needs, easily accessible and ready to go. Kits should include basic survival items and any special items like medications, eyeglasses, or dietary foods.

- **Document Preparation**: Store critical documents in a waterproof and fireproof container. These include identification cards, passports, birth certificates, legal documents like wills, insurance policies, and medical records.

- **Digital Backups**: Consider keeping digital copies of important documents stored in a secure cloud service or on a flash drive within your emergency kit for easy access from anywhere.

Protecting Your Home

While your primary focus should be on the safety of your family, securing your home can prevent further damage and loss.

- **Utilities**: Learn how to shut off gas, water, and electricity in your home to prevent leaks, flooding, or fire. This step should only be taken if there is enough time and doing so does not jeopardize your safety.

- **Secure Loose Items**: Move outdoor furniture, toys, and tools into the house or garage. Secure larger items that cannot be moved with straps or anchors to prevent them from becoming hazardous projectiles.

- **Prevent Water Damage**: If expecting floods, consider sandbagging areas around doors or low windows. Clear drains and gutters to prevent water buildup that could lead to roof damage or interior flooding.

- **Lock Up**: Close and lock all doors and windows. If you have security systems, ensure they are activated before you leave.

Following Official Routes and Guidance

Navigating safely during an evacuation involves more than just knowing the route. Here's how to ensure you follow the best practices:

- **Use Designated Evacuation Routes**: These routes are pre-determined by local authorities to optimize traffic flow and enhance safety. They are often cleared of obstacles and regularly monitored by emergency responders.

- **Avoiding Shortcuts and Alternate Routes**: Stick to the official paths even if they seem congested. Shortcuts may be blocked, unsafe, or unfamiliar, which could delay your evacuation or place you in danger.

- **Road Signs and Instructions**: Pay close attention to road signs, barricades, and directions from emergency personnel. These are placed to guide you safely out of the evacuation zone.

- **GPS and Mapping Tools**: Utilize GPS devices and apps with real-time traffic updates to stay informed about road conditions and any changes in evacuation routes. However, always prioritize directions from emergency officials over GPS guidance.

Staying Informed During the Evacuation

Continuous communication during an evacuation can provide critical updates and instructions that enhance your safety:

- **Emergency Radios**: Keep a battery-powered, solar, or hand-crank radio with you to listen to emergency broadcasts. These devices can be crucial when cellular networks are down or overloaded.

- **Mobile Alerts**: Enable wireless emergency alerts (WEA) on your mobile phone to receive instant notifications from local authorities. These alerts can provide specific instructions and updates about the evolving situation.

- **Emergency Apps**: Consider downloading emergency apps from local or national weather services and disaster response agencies. These apps often offer tailored alerts, evacuation maps, and resource locations.

- **Communication with Family and Friends**: Establish a plan for staying in touch with family members if you get separated. Use text messages or messaging apps, which can be more reliable than voice calls in congested networks.

- **Community Support**: Engage with community safety networks or social media groups that provide localized updates during emergencies. These platforms can offer real-time information and support from nearby residents and authorities.

Post-Evacuation Considerations

Registering at Shelters

Upon reaching a designated safe location or shelter, taking the appropriate steps to register yourself and your group is crucial for several reasons:

- **Accountability**: Registering with the shelter ensures that you are accounted for in the official counts by emergency management officials. This helps in organizing resources and assistance more efficiently.

- **Communication with Loved Ones**: Once registered, shelters can help communicate your safety status to concerned family and friends, relieving them of unnecessary worry and helping to manage the flow of information more reliably.

- **Access to Resources**: Registration often comes with access to essential services such as food, medical care, and sleeping arrangements. Being registered helps the shelter management allocate these resources properly to those in need.

Staying in Safe Locations

Remaining in a safe location until it is officially safe to return home is critical to your safety:

- **Avoiding Further Danger**: Areas impacted by disaster may still pose serious risks such as unstable structures, contaminated water, or live electrical wires. Authorities need time to assess and mitigate these dangers before allowing residents to return.

- **Updates from Authorities**: Stay informed through reliable sources, such as emergency radios or official social media updates, about when it is safe to return. Returning home without official clearance can lead to penalties or getting caught in unsafe conditions.

Maintaining Health and Safety

Staying healthy and maintaining hygiene during an emergency shelter situation involves several considerations:

- **Follow Shelter Rules**: Shelters will have specific guidelines regarding food distribution, sleeping arrangements, and use of facilities. Adhering to these rules helps maintain order and safety for everyone.

- **Personal Hygiene**: In close quarters, the risk of illness can increase. Practice good hygiene by washing hands regularly, using sanitizers, and following any specific health advisories issued by the shelter or health officials.

- **Mental Health**: The stress of an emergency situation can be significant. Utilize support resources often available at shelters, such as counseling services, to manage stress and emotional challenges effectively.

- **Special Needs**: If you or a family member has special medical needs, inform shelter staff immediately. This includes requirements for medications, disability accommodations, or dietary restrictions.

Regular Review and Practice

Regular Evacuation Drills

Consistent practice is key to ensuring an effective evacuation in the event of an emergency. Here's why regular drills are essential:

- **Familiarity with the Evacuation Plan**: Regular drills help every household member become familiar with the evacuation plan, reducing panic and confusion during an actual emergency.

- **Identification of Issues**: Drills allow you to identify and address any practical challenges or obstacles that may arise during evacuation, such as blocked routes or inaccessibility for family members with disabilities.

- **Skill Reinforcement**: Regular practice ensures that essential skills, such as shutting off utilities or packing emergency kits quickly, become second nature to all involved.

Updating Evacuation Plans

Keeping your evacuation plan updated is crucial to its effectiveness, especially when changes occur in your living situation:

- **Changes in Household**: Whether it's the addition of a new baby, elderly relatives moving in, or even pets, updates in household dynamics require adjustments to your evacuation strategies to accommodate everyone's needs.

- **Relocation**: Moving to a new home can drastically change your primary and secondary evacuation routes and meeting points. Reassess and modify your evacuation plan accordingly to fit your new living situation.

- **Community Infrastructure Changes**: Updates to local infrastructure, such as new roads or construction, can affect your planned routes. Keeping abreast of these changes ensures your routes remain viable and safe.

Community Collaboration

Engaging with your local community's safety initiatives can significantly enhance your personal evacuation preparedness:

- **Learning from Others**: Community drills often reveal a range of strategies and considerations that you might not have thought of on your own, allowing you to incorporate these insights into your own plans.

- **Building Relationships**: Regular interaction with neighbors and local emergency managers through community safety programs can build essential relationships that prove invaluable during actual emergencies.

- **Resource Sharing**: Community collaboration can lead to shared resources during an emergency, such like as pooled transportation or collective shelters, increasing the overall effectiveness of community response efforts.

By regularly drilling, updating your plans, and collaborating with your community, you can ensure that your evacuation strategies are robust, flexible, and inclusive, thereby enhancing the safety and preparedness of everyone involved.

Managing Communications During Crises

Effective communication is a cornerstone of crisis management. During an emergency, the ability to disseceive, interpret, and transmit information accurately and quickly can determine the success of response efforts and ultimately save lives. Here's how to manage communications effectively during a crisis:

Establishing a Communication Plan

1. Identification of Key Contacts

Creating an effective communication plan begins with identifying all essential contacts who may need to be reached during a crisis. This comprehensive list should include:

- **Family Members**: Ensure every member of the household has access to contact information for immediate and extended family members, especially those not living in the same household.

- **Emergency Services**: Include numbers for local police, fire departments, hospitals, and any other relevant emergency services.

- **Neighbors**: Maintain a list of contact information for neighbors. In many cases, neighbors can provide immediate assistance faster than emergency services.

- **Utility Companies**: Have the contact details for electricity, water, gas, and internet providers, as these services may need to be reported and managed during a disaster.

Ensure that this list is readily accessible to every member of your household and consider keeping copies in various formats:

- **Digital**: Store in phones, computers, or cloud storage for easy access.

- **Physical**: Keep printed copies in emergency kits, wallets, or other secure, easily accessible locations.

2. Assigning Communication Roles

Delegating specific communication tasks helps streamline the process and ensures that no critical tasks are overlooked during a crisis. Roles might include:

- **Primary Contact**: Responsible for communicating with external entities like emergency services and relaying critical information back to the family or group.

- **Family Coordinator**: Keeps all family members informed of the situation and coordinates with the primary contact.

- **Media Manager**: Manages communications over social media to inform extended family or the public and to receive updates from local authorities.

Training each individual in their respective roles before a crisis ensures everyone knows their responsibilities and can perform them under stress.

3. Choosing Reliable Communication Tools

The effectiveness of a crisis communication plan also heavily depends on the tools available. Different tools will have different uses and reliability under various conditions:

- **Landline Phones**: Often remain operational when cellular networks are overloaded, but may fail in power outages unless equipped with battery backups.

- **Cell Phones**: Widely used but can be unreliable in situations where cellular towers are damaged or overwhelmed. Text messages might be more reliable than voice calls as they use less bandwidth.

- **Satellite Phones**: Provide reliable service in remote areas or when other systems are down, but can be cost-prohibitive.

- **Email and Online Messaging**: Useful if internet access is available, providing a method to communicate detailed information and updates.

- **Radio Communications Systems**: Include CB radios or walkie-talkies that do not rely on cellular networks and can operate over long distances.

Enhancing Communication Effectiveness

1. Redundancy in Communication Systems

To ensure continuous communication capabilities during a crisis, it is crucial to implement redundant systems. This strategy involves setting up multiple communication methods so that if one fails, others can take its place without significant disruption. Key approaches include:

- **Digital and Analog Systems**: Combining modern digital tools like smartphones and traditional analog devices like radios ensures that communication remains possible even if more contemporary networks are compromised.

- **Hard Copies and Digital Data**: Keeping both digital and physical copies of important documents ensures that critical information is accessible even if electronic devices fail.

Implementing these systems involves careful planning and investment in diverse technologies to prepare for various failure scenarios.

2. Regular Testing and Maintenance

Regular testing and maintenance of communication equipment are essential to ensure readiness in a crisis. Steps to ensure operational reliability include:

- **Routine Checks**: Regularly test all communication devices, including emergency radios, cell phones, satellite phones, and internet-based communication tools, to ensure they are working correctly.

- **Battery Management**: Keep all batteries charged, and regularly check their condition. Have a routine for testing and replacing batteries as needed.

- **Backup Power**: Invest in alternative power sources like generators and solar chargers to keep communication devices operational during power outages.

Scheduled maintenance and testing should be part of regular emergency preparedness drills.

3. Clear and Concise Messaging

Effective communication during a crisis depends on the ability to deliver clear and concise messages. Misunderstandings can lead to confusion and inefficiency, potentially escalating the situation. To enhance the clarity of communications:

- **Plain Language**: Use simple, straightforward language that can be easily understood by all recipients. Avoid technical terms and jargon unless absolutely necessary and known to be understood by the audience.

- **Message Structure**: Organize information logically, highlighting the most critical points first. Use bullet points or numbered lists to clarify instructions or important details.

- **Training**: Provide training for all designated communicators in effective communication techniques, including stress communication strategies and the importance of tone and clarity under pressure.

Training and Preparedness

1. Crisis Communication Training

Effective communication during a crisis is crucial and requires specialized skills and preparations. To ensure all team members are prepared:

- **Regular Workshops**: Conduct workshops and training sessions focusing on crisis communication skills, which should include scenario-based practice to handle various types of emergencies.

- **Media Handling**: Train designated spokespersons in dealing with media inquiries to ensure consistent and accurate information dissemination.

- **Empathy Training**: Include sessions on how to communicate empathetically with affected individuals or communities, understanding their distress and providing clear, supportive responses.

Implementing a continuous training program helps maintain a high level of readiness among all team members and ensures that they are equipped to manage communications effectively under stress.

2. Public Awareness Campaigns

Raising awareness about how to communicate during crises can significantly enhance community resilience. Effective strategies include:

- **Community Engagement**: Regularly engage with the community through meetings and workshops to inform them about the communication protocols for various crisis scenarios.

- **Multi-Platform Outreach**: Utilize various communication platforms, such as social media, local newspapers, and community bulletins, to reach different segments of the community.

- **Educational Materials**: Distribute flyers, posters, and digital content that outline key communication channels and steps the public should follow during emergencies.

These campaigns ensure that the community is not only aware of how to receive and interpret information during crises but also knows how to contribute to the safety and efficiency of emergency responses.

3. Simulation Drills

Simulation drills are critical in testing the effectiveness of crisis communication plans and making necessary adjustments based on performance:

- **Realistic Scenarios**: Design simulation drills that mimic potential local emergencies, engaging all relevant stakeholders, including emergency services, local government, and community groups.

- **Feedback Mechanism**: Implement a robust feedback system to gather insights from participants after each drill. This feedback is crucial for identifying strengths and weaknesses in the current communication strategy.

- **Regular Reviews**: Schedule drills regularly and review the communication plan annually to incorporate lessons learned and adjust strategies as needed.

Maintaining Information Flow

1. Central Information Hub

A central information hub is essential for effective communication coordination during crises. Its primary functions should include:

- **Centralized Control**: The hub serves as the primary point for all crisis communication, ensuring messages are consistent and aligned with the latest developments.

- **Information Verification**: It plays a critical role in verifying information before dissemination to prevent the spread of misinformation.

- **Access for All Stakeholders**: Make sure the hub is accessible to all stakeholders, including emergency teams, government agencies, the media, and the public, providing them with timely and accurate information.

Setting up a central information hub involves technology solutions like dedicated software for crisis management, trained personnel who can manage stressful situations effectively, and clear protocols for information flow and escalation.

2. Update Regularly

Regular updates are crucial to maintaining order and trust during a crisis. Effective strategies for regular updates include:

- **Scheduled Briefings**: Establish regular briefing times when updates will be provided to keep all stakeholders informed.

- **Diverse Communication Channels**: Use various communication methods, such as press conferences, social media, email alerts, and local broadcasting, to reach different audiences effectively.

- **Transparency**: Be transparent about the challenges and uncertainties, as well as the steps being taken to manage the situation. This helps in maintaining public trust and managing expectations.

Regular updates not only keep stakeholders informed but also play a vital role in preventing the spread of rumors and anxiety during uncertain times.

3. Feedback Mechanism

A robust feedback mechanism enhances the effectiveness of crisis communication by engaging with the community and stakeholders directly. Key aspects include:

- **Two-Way Communication Channels**: Implement channels like hotlines, social media, and mobile apps that allow stakeholders to give feedback and ask questions.

- **Real-Time Response Team**: Designate a team to monitor these channels and provide real-time responses to ensure that feedback is collected and addressed promptly.

- **Integration with Response Efforts**: Use the feedback to adjust and improve ongoing response efforts. This can involve refining evacuation routes, reallocating resources, or addressing specific concerns raised by the community.

By structuring and maintaining robust communication practices, organizations and families can significantly enhance their resilience and response capabilities during crises, ensuring that all members are informed, prepared, and able to react appropriately as situations evolve.

Book 10: Mindset for Off-Grid Living in Emergency Situations

Embarking on an off-grid lifestyle is a profound shift that extends beyond the mere physical preparations; it is a journey that begins in the mind. Mental preparedness forms the cornerstone of this lifestyle, where resilience and adaptability are not just beneficial traits but essential ones. Understanding and nurturing these qualities can dramatically increase the likelihood of success and fulfillment in an off-grid environment.

Off-grid living offers a way to disconnect from the typical societal structures and demands a high degree of self-sufficiency. However, this disconnection can also lead to significant challenges such as isolation, resource scarcity, and the constant potential for unexpected emergencies. These challenges can test one's mental fortitude, demanding a resilient and adaptable mindset.

Isolation can impact psychological well-being, leading to feelings of loneliness or abandonment in the absence of a traditional community support structure. Resource scarcity, another common issue, requires individuals to manage limited supplies creatively, often under stress, which can strain mental health. Moreover, unexpected emergencies, from natural disasters to medical crises, demand immediate and effective responses, often under extreme pressure.

Developing resilience is thus crucial. It involves more than just enduring; it requires thriving under these conditions. Strategies for building resilience include maintaining robust physical health—regular exercise and a nutritious diet help fortify the body against the stress of off-grid living. Equally important are stress management techniques such as deep breathing exercises, yoga, or even simple regular walks in nature, which help maintain mental equilibrium.

Real-life stories of those who have navigated these challenges successfully can serve as powerful guides. For instance, a family might share their journey of moving to an off-grid homestead, highlighting how they managed resources and dealt with isolation by creating a routine that included both work and leisure, helping them to maintain a balance and prevent burnout.

Emotional preparedness is another critical aspect of off-grid living. Dealing with the emotional highs and lows typical of such a drastic lifestyle change involves having effective coping mechanisms in place. Techniques like mindfulness and meditation can enhance emotional regulation, while establishing a supportive community or network can provide emotional and logistical support.

Adaptability, particularly in times of crisis, transforms challenges into opportunities for growth and learning. Training oneself to think quickly, make decisive decisions, and remain calm under pressure are invaluable skills when living off-grid. Regular practice through simulations of emergency scenarios can help cultivate these skills, ensuring that when real crises occur, one is prepared not just physically but mentally as well.

The importance of practical skills cannot be overstated. Knowing how to perform first aid, respond to emergencies, and possess basic survival skills can significantly boost confidence and reduce anxiety about potential dangers. Learning these skills can be facilitated through community workshops, online courses, or by joining local groups that focus on survival skills training.

Building and maintaining a strong community network is invaluable in off-grid living. Such communities provide emotional support, share resources, and offer advice based on shared experiences, making the off-grid journey less daunting. Engaging with a community can be as simple as participating in local events, initiating projects that benefit all, or using online platforms to connect and share knowledge.

Finally, adopting a long-term perspective on off-grid living encourages sustainability, continual learning, and growth. Setting realistic goals and managing expectations can prevent burnout and frustration. Emphasizing continual learning and adaptation helps individuals remain responsive to changing conditions and emerging challenges, ensuring that the off-grid lifestyle is sustainable and fulfilling over the long term.

In summary, off-grid living is as much about building a resilient and adaptable mind as it is about gathering supplies and skills. It demands a comprehensive approach that combines mental, emotional, and practical readiness, supported by a community that values sustainability, collaboration, and mutual support. Such a preparedness strategy ensures not only survival but a thriving, fulfilling life away from the grid.

Conclusion

Summary of Key Principles for Sustainable, Off-Grid Living

Living off-grid involves more than just disconnecting from municipal services; it encompasses a sustainable lifestyle that reduces one's environmental impact and promotes self-sufficiency. Here are the key principles to consider when planning for sustainable, off-grid living:

1. Energy Self-Sufficiency

- **Renewable Energy Sources**: Utilize solar panels, wind turbines, and micro-hydro systems to generate electricity. These systems should be sized appropriately for your energy needs and geographic location.

- **Energy Storage and Management**: Invest in high-quality batteries to store excess energy and use energy management systems to monitor and control usage efficiently.

2. Water Conservation and Management

- **Rainwater Harvesting**: Install systems to collect, store, and purify rainwater for drinking, cooking, and irrigation.

- **Water Saving Fixtures**: Use low-flow toilets, showerheads, and faucets to minimize water use. Greywater systems can recycle water from baths, sinks, and washing machines for garden irrigation.

3. Waste Reduction and Management

- **Composting Toilets**: Consider the installation of composting toilets as a waterless alternative to traditional plumbing, reducing water usage and producing compost that can be used to enrich the soil.

- **Recycling and Composting**: Separate waste into recyclables, compostables, and trash to minimize landfill waste and repurpose materials and organic waste.

4. Food Independence

- **Permaculture Gardens**: Design and maintain a permaculture garden that supports a variety of edible plants and mimics natural ecosystems to enhance biodiversity.

- **Animal Husbandry**: If space allows, raising chickens, goats, or bees can provide food products such as eggs, milk, and honey.

5. Sustainable Building Materials

- **Eco-Friendly Construction**: Use sustainable, locally sourced materials such as bamboo, reclaimed wood, or straw bales. These materials have lower carbon footprints and are often more suitable for off-grid environments.

- **Efficient Insulation**: Proper insulation is critical to reduce heating and cooling needs, maintaining temperature stability inside the home using minimal energy.

6. Community and Network Support

- **Building Local Networks**: Engage with local communities and form networks that can offer support, trade, and shared resources.

- **Skill Sharing and Education**: Participate in or organize workshops that focus on skills needed for off-grid living, such as solar panel installation, organic gardening, or water filtration.

7. Technology and Innovation

- **Smart Off-Grid Systems**: Utilize smart technology to optimize resource use and automate tasks, from irrigation systems to energy consumption monitoring.

- **Innovative Off-Grid Solutions**: Stay informed about new technologies and approaches that can enhance off-grid living conditions, such as biogas generators or atmospheric water generators.

By integrating these principles, individuals can create a resilient, sustainable lifestyle that is not only beneficial for the environment but also provides a sense of independence and fulfillment. Off-grid living requires careful planning, commitment, and adaptation, but with the right approach, it can offer a rewarding way of life.

Starting Your Off-Grid Journey

Embarking on an off-grid journey is a transformative commitment that marries sustainability with a profound connection to nature. This guide is your call to action, a beacon for those ready to step away from the grid, reduce their ecological footprint, and embrace a life more attuned to the environment.

Begin by immersing yourself in the world of off-grid living. Dive into books, documentaries, and firsthand accounts from those who've paved the path before you. Understand the core components—from harnessing rainwater to managing waste and generating your own power. This knowledge is your foundation, the bedrock upon which you'll build your new life.

Assess your needs critically. Reflect on your current lifestyle—what you can adapt, what you can't live without. Consider your climate, your location, the land. Each factor weaves into your decision, influencing how you'll shape your off-grid home.

Start small. You don't need to leap into complete self-sufficiency overnight. Test the waters with manageable projects: a small vegetable garden, perhaps, or a simple rainwater collection system. These initial steps are both practical and profoundly educational, teaching you the rhythms of nature and the basics of self-reliance.

As you grow more confident, expand your skills. Off-grid living demands versatility—gardening, carpentry, mechanical repairs, renewable energy installation. Seek out workshops and courses. Practice. Learn. Grow.

Financial planning cannot be overlooked. Chart the costs of setting up and sustaining your off-grid home. Save diligently, explore grants for renewable energy, and budget with an eye toward sustainability. This financial foresight ensures your transition is grounded, not in whimsy, but in practicality.

Choosing the right location is crucial. Your land should support your dreams, offering access to water, space for crops, and compliance with local zoning laws. It should be a canvas on which you'll paint your new life.

Community is vital. Off-grid living thrives on shared knowledge and support. Forge connections with like-minded individuals, engage in local projects, and build networks that will sustain you just as much as your vegetable garden or solar panels.

At every decision point, choose sustainability. Opt for renewable resources, recycle and reuse, and minimize your ecological impact. Your off-grid home should be a testament to a lifestyle that respects and protects the planet.

Prepare for challenges—they are inevitable. Whether it's a sudden repair or an adjustment to off-grid living, approach each with resilience and creativity. Each obstacle is an opportunity to learn and to grow.

When you're ready, take the leap. Transitioning to off-grid living is not just changing where you live, it's changing how you live. It's an ongoing journey of learning and adaptation, filled with both challenges and immense rewards.

Imagine your life, not as a series of days dictated by the grid, but as a tapestry of experiences woven from self-reliance and sustainability. This is more than a change of scenery; it's a new way of seeing the world and your place within it. Embrace it, and let the adventure begin.

Made in the USA
Monee, IL
02 October 2024

67095015R00096